BULLETIN OF THE JOHN RYLANDS LIBRARY

Imaging Heritage Science at the John Rylands Research Institute and Library, University of Manchester

Guest Editors: Stefan Hanß and James Robinson

BULLETIN OF THE JOHN RYLANDS LIBRARY

ISSN 2054-9318 (Print)
ISSN 2054-9326 (Online)
ISBN 9781807070106

Established in 1903

Published by Manchester University Press and the University of Manchester Library
Manchester University Press, 176 Waterloo Place, Manchester, M13 9GP, UK
Email: mup@manchester.ac.uk
Web address: www.manchesteruniversitypress.co.uk

University of Manchester Library, Oxford Road, Manchester, M13 9PP, UK
Tel: +44 (0)161 275 3751
Web address: www.library.manchester.ac.uk

The publication of the *Bulletin of the John Rylands Library* is made possible by funding from The University of Manchester Library.

Members of the Editorial Board 2025

Chair: David Matthews, University of Manchester
Editors: Fred Schurink, Rachel Winchcombe and Huw Twiston Davies, University of Manchester
Editorial Assistant: Emma Nelson, University of Manchester

Editorial Board

Guyda Armstrong, University of Manchester
Paul Fouracre, University of Manchester
Roy Gibson, Durham University
David Law, University of Manchester
Janette Martin, John Rylands Library, University of Manchester
John Morgan, University of Bristol
Walter Pohl, Austrian Historical Research Institute
Lynda Pratt, University of Nottingham
Ingrid Rembold, University of Manchester
Julianne Simpson, Chetham's Library
Carsten Timmermann, University of Manchester
Huw Twiston Davies, University of Manchester
Jack Webb, University of Manchester

Subscriptions

An electronic version of this issue is available to read online under a CC BY-NC-ND licence, at
https://www.manchesterhive.com/view/journals/bjrl/bjrl-overview.xml

Print editions for issues 96.2 and onwards are available to purchase at
https://manchesteruniversitypress.co.uk

EU authorised representative for GPSR:
Easy Access System Europe, Mustamäe tee 50, 10621 Tallinn, Estonia
gpsr.requests@easproject.com

BULLETIN OF THE JOHN RYLANDS LIBRARY

VOLUME 101 NUMBER 2 AUTUMN 2025

CONTENTS

Imaging Heritage Science Initiatives at the John Rylands Research Institute and Library *Stefan Hanß and James Robinson*	1
'Fraught with Possibilities of World-Wide Benefit': Towards a History of Photography at the John Rylands Library *Elizabeth Gow, John Hodgson, Tony Richards and James Robinson*	15
Imaging Heritage Featherwork: A New Methodology for the Study of Feather Artefacts *Stefan Hanß, James Robinson and Tony Richards*	35
Sri Guru Granth Sahib: Collaboration, Digitisation, Heritage and the Legacy of Colonial Collections *Gurtek Singh and James Robinson*	55
Imaging the Gaster Jewish Amulets in the John Rylands Research Institute and Library *Philip Alexander, James Robinson, Elizabeth Evans, Amin Garbout, Jo Castle, Tony Richards and Ira Rabin*	73
The *Apocalypse* and *Biblia pauperum* Blockbooks Bound by Johannes Richenbach in 1467 *Stephen Mossman and Edward Potten*	91
The Trier Psalter-Hymnal (Manchester, John Rylands Research Institute and Library, MS Lat. 116): Palette and Pigments *Richard Gameson and Andrew Beeby*	109

Imaging Heritage Science Initiatives at the John Rylands Research Institute and Library

STEFAN HANß, UNIVERSITY OF MANCHESTER
JAMES ROBINSON, UNIVERSITY OF MANCHESTER

Imaging is transforming the ways we conduct research on cultural heritage; the ways heritage artefacts are preserved, examined, understood, displayed and made relevant for various communities and their specific needs across the world; as well as how heritage communities past and present get involved. The Imaging Laboratory of the John Rylands Research Institute and Library at the University of Manchester, this special issue shows, is at the very forefront of such developments. The history of photography at the John Rylands Research Institute and Library covers a period of nearly 120 years, evidencing an impressive transition from some of the earliest heritage photography in the library's early days to the most cutting-edge imaging heritage science projects today.

For decades, Rylands imaging photographers and scientists – in collaboration with academics, communities and community researchers, curators and conservators – have established, developed and enhanced new applications for imaging technologies that have resulted in both significant methodological advancements and the enhancement of best practice in imaging heritage science. Such initiatives have channelled important relationships with academia and communities, curators, conservators, heritage practitioners and technicians, as well as across collections and laboratories within and beyond Manchester. These partnerships inform our understanding of imaging heritage science as a practice of relationship-building. Imaging heritage science research at the John Rylands Research Institute and Library has established strong partnerships that broker methodologically innovative, new collaborations that result in interdisciplinary research and groundbreaking methodological advancements. These Rylands experiences, in turn, are reshaping the way that we think about imaging as a scientific research method and practice in heritage studies.

This *Bulletin of the John Rylands Library* special issue, available open access via University of Manchester Press, puts imaging heritage science centre stage. Besides this thematic focus, there is an emphasis on cultures of research innovation. This special issue shows what can be achieved when open-minded people with unique expertise and a passion for their work and curiosity for new conversations join forces: we can create a platform to explore the innovation potential of pathbreaking new methodologies, an inclusive platform for everyone to thrive. This special issue aims to showcase some of the pathbreaking interdisciplinary academic projects on imaging heritage science at the University of Manchester and to highlight the cutting-edge work and collaborations of the John Rylands Research Institute and Library Imaging Laboratory. We hope to broaden the international visibility of our imaging heritage science projects by producing one standard reference – this present special issue – that captures, at least to some extent, the breadth, width and depth of the current work of the Imaging Lab and their academic and community research partners. Most importantly, this special issue presents the crucial interventions of Rylands imaging projects in wider heritage science research.

This essay introduces the agenda, content and wider interventions of the special issue. It is worth stressing that this is an introductory essay rather than an academic introduction in the strict sense.[1] We present a set of reflections based on personal experiences: James has worked for decades as heritage photographer and imaging scientist and is now Imaging Manager at the library, and Stefan has worked as a researching and teaching academic with the library for years, and for several years as Deputy Director and Scientific Lead of the John Rylands Research Institute. We start by outlining our stance on imaging heritage science, arguing for the embeddedness of imaging research in wider heritage science and emphasising the need to consider imaging heritage science as a method rather than a technique. In this first section, we also stress the significance of individual expertise, experience and creativity on the one hand, and collaboration with academics and community researchers on the other, in order to fulfil the potential of imaging heritage research to broker new, valuable relationships. We also introduce the Rylands Imaging Lab – the team, their equipment and experience, as well as contexts and research environment – situating imaging heritage science initiatives more broadly within local dynamics and international heritage research. In a second part, we highlight this special issue's content and key interventions, and stress the overlapping themes and some of the unique insights of the respective articles.

Imaging Heritage Science: Methods and Relationships

Imaging is a widely used technique of image creation and data processing, which is expansive both in terms of application and range of technology. The fields of medicine, archaeology, materials sciences, heritage science and remote sensing – to name just a few – have embraced imaging, most broadly defined as 'the technique or practice of creating images of otherwise invisible aspects of an object'.[2] Although this definition comes with the advantage of a wider perspective, we propose – for the purpose of this special issue – a definition of advanced imaging as embedded in heritage science, using the term 'imaging heritage science'. Most importantly, this shifts the focus of attention from techniques towards methods, and prompts more reflective engagement with advanced imaging of cultural heritage as a practice of relationship-building. As a consequence, some contributions of this special issue discuss more classical advanced imaging technologies alongside wider scientific heritage analysis that has been applied to contextualise particular heritage artefacts. The definition of advanced imaging as imaging heritage science, moreover, calls for a personalisation of discourse. Put differently, as a discipline we must learn to talk at times less about laboratories, so as to give more space to celebrate the people and collaborations that enable such research to thrive.

We consider imaging heritage science less a technique than a *method*. Imaging thus goes beyond digitisation – here defined as 'the representation of an analogue original in a digital format' – and is first of all a form of enquiry: a mode of thinking about the ways that making the invisible visible through imaging technologies will foster, shape and enhance engagement with the material artefact in appropriate ways.[3] As a method, imaging heritage science does not apply new or different technologies for the sake of it; we seek to use and, crucially, advance enhanced imaging to deepen understandings of and widen engagement with heritage artefacts and communities. This definition of advanced imaging as imaging heritage science is in line with more recent process- and relations-driven understandings of heritage, as opposed to product-focused definitions.[4] Considering imaging a method, we argue, is required to exploit fully the opportunities that advanced imaging offers to unlock and enquire into historical heritage artefacts, and for shaping the trajectory of future heritage studies more generally.[5]

The definition of imaging heritage science as a problem-driven mode of enquiry invites social reflections on contemporary imaging work. As Siegfried Kracauer once aptly put it, ours is a time characterised by a 'flood of photos': 'never before has an age been so informed about itself, if being informed means having an image of objects that resembles them in a photographic sense', and yet 'never before has a period known so little about itself'.[6] Kracauer wrote in 1927, but his reflections on the extent to which a multitude of images can inhibit rather than inform or encourage critical engagement – 'the likeness that the image bears to it effaces the contours of the object's "history"', Kracauer wrote, generating instead 'an indifference toward what the things mean' – could hardly be more relevant to the present age.[7] Today, we experience the extent to which the politics of the image can reshape the image of politics, *sensu* Walter Benjamin.[8] Building on Kracauer's observations, Theodor Adorno insisted that 'omnipresent images' not only document but themselves reproduce the problematic logics that drive mass consumption and mass culture. As Adorno stated, 'the sheer quantity of the material processed ... has become quite incommensurable with the horizons of individual experience'.[9] Scholars of colonialism and slavery in particular have expressed very similar concerns over the ways that the 'logic of quantification' ingrained in mass digitisation 'can stand in the way of centring the experiences of the communities who lived under colonialism and slavery'.[10] The cultural repetition of violence, power imbalances and biases, as well as issues of legitimacy that come with the unreflective mass imaging of sensitive materials, have been clearly named: 'the institutional drive to reproduce the excessive scale of the colonial project as big data', Temi Odumosu warns, 'enacts its own forms of erasure'.[11] Archaeologists have therefore coined the terms 'digital colonisation' and 'digital colonialism' to emphasise that mass digital reproduction, big data and patronising debates about so-called digital repatriation reproduce epistemic violence – disputed inheritance, denied ownership and refused self-determination – anchored in colonial thought, imperial practice and the institution of slavery.[12] Yet, as Kracauer reminds us, 'it would not have to be this way', and situating 'access and permission, ownership, decontextualization of heritage, and the digital economy that comes with digitizing cultural heritage' within 'an inclusive ethical framework', as Bijan Rouhani demands, is crucial.[13] Acknowledging vulnerability, violence and community needs, we stress, is of similar significance. Once imaging heritage science does so, we can transcend debates on epistemics to embrace the opportunities that imaging encapsulates as a research practice that shapes relationships.

Imaging heritage science, we argue, is a *practice* to build inclusive, creative, productive and transformative relationships. A culture of trust, appreciation, listening and collaboration is key to build cutting-edge imaging heritage research. We therefore acknowledge the unique environment in which Rylands Imaging Lab activities are situated, in particular the close working relationships with academic and community researchers, with the library's conservators and curators as well as with other professionals in the field, to exploit knowledge and experience with collections and technical ability to the benefits of research and methodological advancements. Rylands imaging heritage science relationships are anchored in a wider environment that reflects on both the unique expertise of library curators and conservators as well as the unique chronological, cultural, geographic, linguistic and material depth of our Special Collections. The University of Manchester Library's Special Collections of rare books, manuscripts, archives, maps and visual collections, among others, are internationally significant and were Designated by Arts Council England in 2005. They are ranked in the top five University Special Collections in the world.[14]

The John Rylands Research Institute offers a dynamic and thriving research environment, brokering innovative connections and groundbreaking research with academics at Manchester and beyond. The Research Institute has proven to be transformative in connecting research excellence with methodological innovations, labs and academics, people, ideas and external funding. Imaging heritage science research at the library benefits massively from the Research Institute's strategic priority to grow scientific and digital approaches to artefacts. The Research Institute generates major external research and infrastructure grants and spearheads high-profile international partnerships in heritage science, materials sciences and digital humanities, and with international partner research institutes, laboratories and libraries, as well as with creative industry and community-led research. The Research Institute has also proven key in brokering new heritage science connections with laboratories at Manchester and beyond, to join forces with the university's unique landscape of internationally leading lab facilities. The Research Institute's wide fellowship programme, moreover, ensures constant groundbreaking collections-based research activities that translate into wider grant applications and international as well as interdisciplinary visibility, with many projects comprising cutting-edge imaging, scientific and AI components. This wider environment of strategic and sustained investment into transformative collections-based research allows imaging heritage science to thrive. The visibility and impact of imaging heritage science benefits from the *Bulletin of the John Rylands Library*, which provides a platform to distribute research results, and the new advanced image viewer (Manchester Digital Collections), developed in collaboration with the University of Cambridge, which is transforming access to high-resolution images and rich collection metadata.

The Rylands Imaging Lab Team: Expertise, Experiences, Equipment

Considering imaging heritage science as a practice of relationship-building means acknowledging the unfolding and relational character of analysis. Differently put, we urge a move of attention in imaging heritage science discourse from labs to people, resulting in a stronger and more explicit acknowledgment of the individual experiences and expertise that imaging practitioners bring to collaborations. After all, it is the unique mix of individuals and collaborations that drive innovation potential. Embedding advanced heritage imaging into a broader scientific environment of heritage analysis allows for a redefinition of the benefits of and relationships between photography, advanced imaging, conservation and heritage sciences, both to bridge established disciplinary divides and to reconsider the unique expertise that each single imaging practitioner brings to the debate.[15] The Rylands imaging team has developed over the past eighteen years from one part-time project photographer and a photographer from another former unit within the university, a manager and an assistant to a team of seven members: Jo Castle (Senior Photographer), Amber Greenall-Heffernan (Imaging Assistant), Tony Richards (Senior Photographer), Lisa Risbec (Imaging Assistant), James Robinson (Imaging Manager), Samuel Simpson (Photographer) and Holly Staniforth (Project Photographer). It is their expertise and experience that drives the team's innovation potential, efficiency and creativity. As the invisibility of photographers and technicians – like curators, conservators and translators – is ingrained in the presentation of research, it is time to celebrate, foreground and name the people that make success and innovation happen. It is the unique mix of object biographies and the biographies of imaging scientists and collaborating academics, community researchers and heritage practitioners that widens interpretations and advances our methodological toolkit and understandings.[16]

Rylands imaging scientists advance state-of-the-art technology to inform our understanding of cultural heritage, responding to research questions posed by academics at Manchester and beyond, in close collaboration with conservators and curators at the library, and with local, national and international partners. The beginnings of advanced imaging at the Rylands date back to 2012, when team members removed the infrared cut-off filter from an out-of-warranty Phase One P45+ and began experimenting with LED light panels with a range from ultraviolet 365 nm through the visible spectrum to infrared 840 nm. Such initiatives underwent further testing with the University of Manchester's Photon Science Institute and Department of Computer Science. The team closely collaborated with Mike Toth (R. B. Toth Associates) and Bill Christens-Barry (Equipoise Imaging) to develop multispectral imaging (MSI) capabilities over the years, which evolved into a test case for the development and commercialisation of hardware via the Phase One Rainbow system. The John Rylands Research Institute and Library now hosts a full studio of Phase One IQ4 digital systems and continues to invest into the newest technology. This includes major infrastructure and research project grant initiatives led by the John Rylands Research Institute that helped upgrade, replace and further advance imaging equipment, as well as the Research Institute's high-profile international partnerships with academic and scientific partner institutions and labs worldwide.[17] The Imaging Lab has also recently purchased a Selene Photometric Stereo System from Factum Foundation, making the John Rylands Research Institute and Library one of three UK institutions to have such a system. These investments and ongoing and developing research grant and infrastructure funding initiatives, nationally and internationally, are essential to future-proof one of the leading heritage imaging units worldwide.

The move of the Imaging Lab into new facilities in 2025 has created an additional momentum for groundbreaking imaging research. Although the previous studio space has long been used as photographic room (Figure 1; see also Figure 2 in the article '"Fraught with Possibilities of World-Wide Benefit": Towards a History of Photography at the John Rylands Library', in this special issue), technical

Figure 1 The French Revolution Room's transformation over time: glass-plate negative image of the room (*c.* 1960).

Figure 2 The French Revolution Room's transformation over time: the collection storage room in 2023.

advancements and new knowledge of collection care have rendered this space unsuitable for a long time. Access conditions posed problems around the use and transport of sensitive or particularly heavy collection items, as did fluctuating environmental conditions. Several alternative spaces had been in the conversation for years, among them the Dante Room, which is now used for storage, or the cellar space, the typical dwelling of digitisation units across the globe. For a long time, the team had their eye on the French Revolution Room, which housed a substantial collection of newspapers, periodicals and books published at the time of the French Revolution and up to the Restoration and the Bourbons, including 15,000 broadsides, proclamations and bulletins, as well as contemporary periodicals and newspapers (Figure 2).[18] As part of the Rylands 'Next Chapter' project's architectural works in 2024–25, the French Revolution Room has been now transformed into new imaging lab facilities (Figure 3).

The development of the new laboratory brings major advantages. Most importantly, the team is now working together in one dedicated space. Full climate control allows the laboratory to match conditions in the secure stores, meaning items come under no environmental stresses. The space is fully modular and movable, with bays divided by blackout curtains to accommodate larger items and facilitate more collaborative work with researchers, including in-person or hybrid research seminars, workshops and training or teaching activities. The team can now also use a new 86-inch display with conferencing capabilities, and a new doorway and ramp grant direct access to a large lift and storage areas.

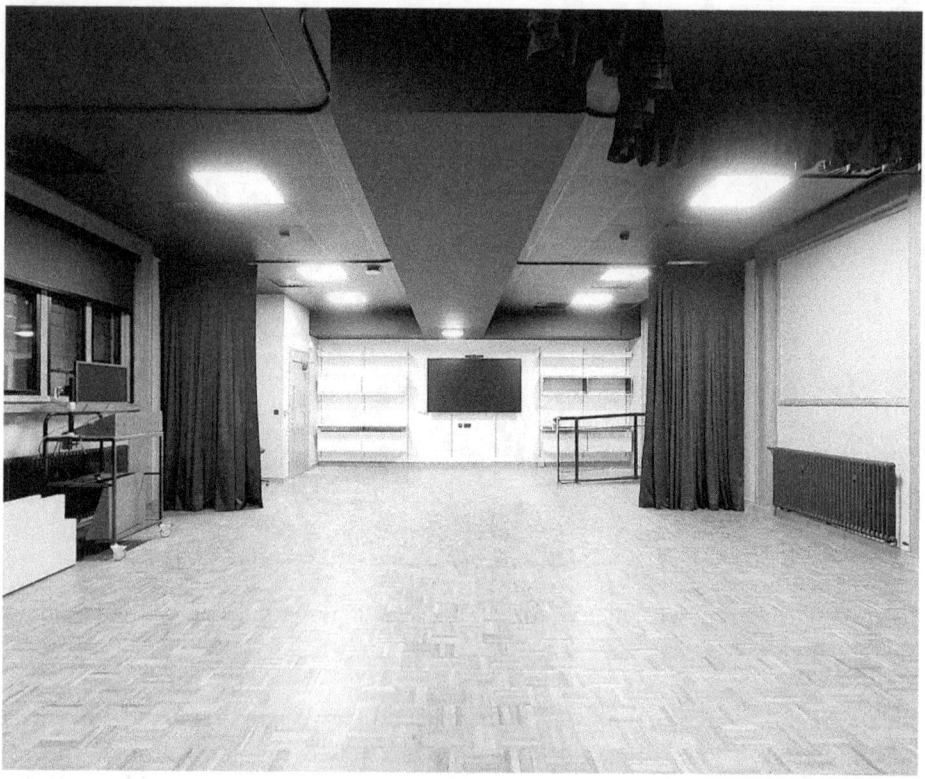

Figure 3 The French Revolution Room's transformation over time: new Imaging Lab facilities at the John Rylands Research Institute and Library in 2025.

The advantages of relocating to these new lab facilities are already evident, resulting in significant research opportunities. For the first time, one of the most substantial and historically significant items in the Rylands collections – French MS 1*, the 1546 Mappemonde by French cartographer Pierre Desceliers (born c. 1537) – can be comprehensively examined.[19] Due to nineteenth-century interventions, the map has remained largely inaccessible for detailed research and scientific analysis. Mounted on two 2.7-metre-long oak poles, rolled and stored within a flight case, the map's sheer size and fragility have historically made transportation and handling highly challenging. The newly established laboratory, however, has already facilitated a week-long, collaborative project on the map. During this initial analysis within the Imaging Lab, the Collection Care team has undertaken detailed condition assessments, while the Imaging team has conducted high-resolution imaging, 3D photogrammetry, macrophotography and multispectral imaging (Figures 4 and 5). This marks a critical first step towards implementing preventative conservation measures and enabling more in-depth scientific and interpretative analysis of this unique artefact. Ultimately, the initiative aims to

Figure 4 James Robinson (Imaging Manager), Tony Richards (Senior Photographer) and Samuel Simpson (Photographer) analysing French MS 1* in the new Rylands Imaging Lab facilities in 2025. Photograph taken by Holly Staniforth.

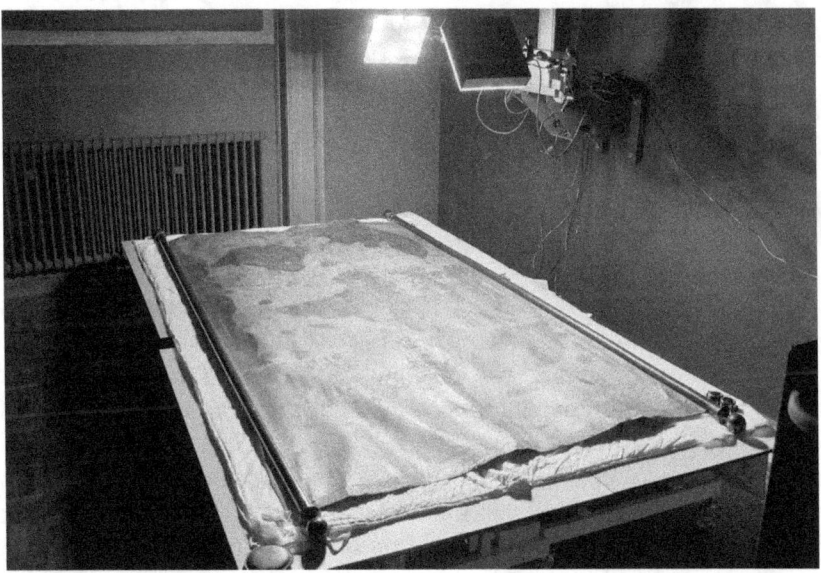

Figure 5 Multispectral imaging conducted on Pierre Desceliers, Mappemonde, 1546 (JRRIL, French MS 1*) in the new Rylands Imaging Lab.

generate a comprehensive dataset that will facilitate and enhance future, large-scale research.

In short, the definition of advanced imaging heritage initiatives as imaging heritage science – and thus as part of the wider scientific enquiry of cultural artefacts – comes with three major transitions: we turn from techniques to methods and methodological advancements; from digitisation products to the practice of relationship-building; and from labs to people. At the John Rylands Research Institute and Library, this approach has resulted in cutting-edge projects. Imaging scientists have made the invisible visible: they have recovered palimpsests, faded or erased texts, analysed pigments and revealed the content of hidden amulets. Rylands imaging heritage science is spearheading ethical debates about digitisation and the legacies of colonial practice and epistemic violence in cultural institutions. The John Rylands Research Institute is also hosting a number of projects that bring computer vision, artificial intelligence and digital microscopy to imaging approaches. We provide cutting-edge multispectral imaging analysis, macrophotography, focus stacking and 3D visualisation and modelling of heritage artefacts of the highest international level of quality, and team members have developed new methodologies of advanced imaging enhancing research, such as on historical watermarks. In partnership with academics and other collections and institutions, the Rylands Imaging Lab has conducted research on a wide range of materials, both those housed at the library as well as those on loan from other institutions, including a mummy cartonnage from ancient Egypt and a feather fan from seventeenth-century Brazil. Rylands imaging scientists combine expertise in photography and historical photography with a unique commitment to the most cutting-edge technologies which has resulted in creative and problem-driven approaches to solution-finding that help transform technologies and advance methods. The team's willingness to share knowledge and their commitment to enhance best practice in the field creates considerable potential for the wider field of imaging heritage science.[20]

Special Issue Content and Interventions

This issue's content reflects on the rich chronological, cultural, geographical, material and religious depth of Rylands collections and imaging traditions, while introducing some – but not all – of the imaging techniques that are used to generate new insights and partnerships. The contributions seek to speak to a general audience and specialists and imaging experts alike. In the first article, Elizabeth Gow, John Hodgson, Tony Richards and James Robinson situate today's cutting-edge Rylands imaging heritage science initiatives in the wider history of photography at the library. Since 1910, the Rylands has been at the forefront of heritage photography, which has been a key contribution to widening access to collections, facilitating research and enhancing the library's prestige and visibility. Building on uniquely dense and rich archival documentation, this article historicises the practice of imaging. It showcases the transition from heritage photography to imaging heritage science and problematises the (in)visibility of imaging specialists.

Stefan Hanß, James Robinson and Tony Richards then present a discussion of new Rylands imaging research conducted on heritage featherwork at the Fitzwilliam Museum, University of Cambridge. This article showcases some of the exciting collaborations across collections, disciplines and institutions, as well as the insights that Manchester imaging science expertise can bring to research on heritage artefacts beyond library contexts. Building on existing scholarship by biologists, conservators and ornithologists who evidenced spectral curve responses of feathers to different wavelengths, as well as work with feather artisans and artist practitioners, the project's principal investigator at the University of Manchester has partnered with the Rylands Imaging Lab to connect UK-wide leading imaging technology with heritage collections across the country to advance the analysis of historic featherwork. The article discusses the use of multispectral imaging, focus stacking, digital microscopy, transmitted light imaging and 3D views generated by reflectance transformation imaging (RTI)/polynomial texture mapping (PTM) as new methods for the study of historic featherwork. Such advanced imaging research holds surprising new insights in terms of feather and species identification, pigment and dye analysis and the study of historic and endangered crafts, as well as for the conservation and display of such fragile and critically endangered heritage artefacts. The article calls for new collaborations with Indigenous communities in the study of featherwork, and highlights transformative research partnerships across collections and communities.

Gurtek Singh and James Robinson then present groundbreaking community-led research on the Sri Guru Granth Sahib at the John Rylands Research Institute and Library. This is an exemplary study of colonial collections and a major contribution to the field of decolonial and postcolonial studies. The article presents a new methodology of how to digitise with cultural awareness and in response to the needs of communities, paving the way to new community-led imaging heritage research. This article will be a cornerstone in future debates on the digitisation of looted colonial heritage and the ethics of imaging heritage science. This contribution sheds new light on what it means to work not with texts or manuscripts but with The Guru, who is considered a living being, and with the communities to which The Guru belongs. This article's most crucial methodological intervention is the authors' insistence on the value of respect and mutual learning in community-led imaging heritage science.

Philip Alexander, James Robinson, Zsófia Buda, Elizabeth Evans, Amin Garbout, Jo Castle, Tony Richards and Ira Rabin then present digital imaging analysis conducted on the Rylands collection of Jewish amulets. Different techniques have been used, including 3D photogrammetry, XRF analysis and X-ray computed tomography, to study the Rylands Gaster Hebrew amulets, and here in particular the sealed Gaster Amulet 34. This was the first time that Manchester Special Collections materials have been studied at the National Centre for X-Ray Computed Tomography (NXCT) to reveal the sealed content of heritage artefacts.

The article thus exemplifies the extent to which imaging heritage science initiatives can broker new sector-wide partnerships across laboratories, while also problematising the challenges that come with such new research.

Stephen Mossman and Edward Potten then present novel research on some of the earliest known European blockbooks. The article is based on a new method of analysis of historical watermarks developed by Rylands imaging specialist Tony Richards, who combined transmitted and reflected light images with innovative digital data processing tools. Such analysis has been brought into a conversation with further materials scientific research on fifteenth-century Richenbach blockbooks held at the John Rylands Research Institute and Library. It includes digital microscopy and X-ray fluorescence analysis of inks developed and conducted by Ira Rabin (Federal Institute for Materials Research and Testing Berlin and former Visiting Professor and current Honorary Fellow at the John Rylands Research Institute) and Stefan Hanß (University of Manchester), who also leads and further develops the University of Manchester History Department's collection of digital microscopes – the largest collection of digital microscopes at any faculty of humanities in the UK. This article exemplifies the stunning new insights researchers may gain when heritage science and methodological innovation in advanced imaging is brought to well-established debates such as dating or provenance.

In the final article, Richard Gameson and Andrew Beeby discuss the colour palette of the Trier psalter-hymnal held at the John Rylands Research Institute and Library. While this article offers rare insights into the elemental composition of pigments in Carolingian manuscripts, it also showcases new collaborations between the Rylands Imaging Lab, the John Rylands Research Institute, Team Pigment at Durham University and the Federal Institute for Materials Research and Testing in Berlin (BAM). Pigment and ink analysis of the Trier psalter-hymnal has been conducted by Team Pigment (Durham) and Ira Rabin (BAM) using reflectance spectroscopy, Raman spectroscopy, X-ray fluorescence spectroscopy, multispectral imaging and photomicroscopy. This, again, highlights the extent to which advanced imaging is embedded in a wider culture of scientific materials analysis of cultural heritage.

All these articles present several key interventions and call for actions. The focus on methods and, more importantly, methodological advancement and creative interdisciplinary collaborations to drive new approaches is one of them. Another focus emerges from these articles' reflections on collaborations. Advanced heritage imaging is not just about the delivery of results; it is a form of conducting research, a practice that is anchored in relationships. It is for this reason that we call for the use of imaging heritage science to shape respectful relationships of trust and mutual learning across disciplines, institutions and communities. Another call for action is to widen our understanding and acknowledgement of heritage practitioners and their diverse backgrounds, including community researchers, photographers and technicians. It is their different forms of expertise that drive innovative collaborations and new methods. These articles also evidence the need for collaboration across communities, countries, disciplines and institutions. As advanced imaging works with image data collection and processing, the articles also repeat the call for FAIR data and data processing: data must be findable, accessible, interoperable and reusable. Such interventions will help the sector as it faces challenges around environmental sustainability, artificial intelligence, funding cuts, technology developments and cyber security as well as data dependency.[21] This second issue of the 101st annual run of *The Bulletin of the John Rylands Library* will, we hope, provide food for thought for the development of imaging heritage science in the next 100 years to come.

Notes

1. Footnotes have therefore been condensed for efficiency.
2. W. H. Börner, 'Imaging', in S. L. López Varela and J. Thomas (eds), *The Encyclopedia of Archaeological Sciences* (Chichester: Wiley, 2019), https://doi.org/10.1002/9781119188230.saseas0317.
3. A. Cullingford, 'Digitisation and Digital Libraries in Special Collections', in *The Special Collections Handbook* (London: Facet, 3rd edn, 2022), p. 146.
4. B. Rouhani, 'Ethically Digital: Contested Cultural Heritage in Digital Collections', *Studies in Digital Heritage*, 7:1 (2023), 1–16.
5. M. Terras, 'Implementing Advanced Digital Imaging Research in Cultural Heritage: Building Relationships between Conservators and Computational Imaging Scientists', in Alberto Campagnolo (ed.), *Book Conservation and Digitization: The Challenges of Dialogue and Collaboration* (London: Arc Humanities Press, 2020), pp. 217–31.
6. S. Kracauer, 'Photography', first published 1927, trans. T. Y. Levin, *Critical Inquiry*, 19:3 (1993), 432.
7. *Ibid.*
8. W. Benjamin, 'Das Kunstwerk im Zeitalter seiner technischen Reproduzierbarkeit', in W. Benjamin, *Gesammelte Schriften*, vol. I, 2, eds R. Tiedemann and H. Schweppenhäuser (Frankfurt/Main: Suhrkamp, 1980), pp. 431–69.
9. T. Adorno, *Minima Moralia: Reflections from Damaged Life*, first published 1951, trans. E. F. Jephcott (London: Verso, 2005), p. 140.
10. D. Agostinho, 'Archival Encounters: Rethinking Access and Care in Digital Colonial Archives', *Archival Science*, 19:2 (2019), 157.
11. T. Odumosu, 'The Crying Child: On Colonial Archives, Digitization, and Ethics of Care in the Cultural Commons', *Current Anthropology*, 61:22 (2020), S294.
12. K. L. Krupa and K. T. Grimm, 'Digital Repatriation as a Decolonizing Practice in the Archaeological Archive', *Across the Disciplines*, 18:1–2 (2021), 47–58; E. C. Kansa, 'The Great Digital Lost and Found', *Conservation Perspectives*, 37:2 (2022), 4–9; Pinar Oruç, 'Rethinking Who "Keeps" Heritage: 3D Technology, Repatriation and Copyright', *GRUR International*, 71:12 (2022), 1138–46; Rouhani, 'Ethically Digital'.
13. Kracauer, 'Photography', 432; Rouhani, 'Ethically Digital'.
14. 'Special Collections', https://www.library.manchester.ac.uk/rylands/special-collections/ [accessed 13 February 2025].
15. Terras, 'Implementing Advanced Digital Imaging Research in Cultural Heritage'.

16 A. Appadurai, *The Social Life of Things: Commodities in Cultural Perspective* (Cambridge: Cambridge University Press, 1986); L. Venuti, *The Translator's Invisibility: A History of Translation* (London: Routledge, 1995); F. Galeazzi, '3-D Virtual Replicas and Simulations of the Past: "Real" or "Fake" Representations?', *Current Anthropology*, 59:3 (2018) 265–78.

17 See, for example, the AHRC Capabilities of Collections grant to fund 'Increasing capability for collections research at the University of Manchester' (2021, £778,351), led by Professor Hannah Barker, former director of the John Rylands Research Institute, which funded core multispectral imaging, or the more recent collaboration with the German Federal Institute for Materials Research and Testing (BAM) Berlin.

18 'French Revolution Collection', https://www.library.manchester.ac.uk/rylands/special-collections/a-to-z/detail/?mms_id=992983876727401631 [accessed 13 February 2025].

19 C. A. Burland, 'A Note on the Desceliers' Mappemonde of 1546 in the John Rylands Library', *Bulletin of the John Rylands Library*, 33:2 (1951), 237–41.

20 C. Jones, C. Duffy, A. Gibson and M. Terras, 'Understanding Multispectral Imaging of Cultural Heritage: Determining Best Practice in MSI Analysis of Historical Artefacts', *Journal of Cultural Heritage*, 45 (2020), 339–50.

21 See, for example, 'Learning Lessons from the Cyber-Attack: British Library Cyber Incident Review. 8 March 2024', https://www.bl.uk/home/british-library-cyber-incident-review-8-march-2024.pdf [accessed 13 February 2025].

Manchester University Press

'Fraught with Possibilities of World-Wide Benefit': Towards a History of Photography at the John Rylands Library

ELIZABETH GOW, UNIVERSITY OF MANCHESTER
JOHN HODGSON, UNIVERSITY OF MANCHESTER
TONY RICHARDS, UNIVERSITY OF MANCHESTER
JAMES ROBINSON, UNIVERSITY OF MANCHESTER

Abstract

This article contextualises contemporary heritage imaging science at the John Rylands Library by exploring the twentieth-century development of its photographic services. Taking an archives-led approach, we argue that the core purpose of photography was to support research, primarily by providing access to textual content. We also consider the status and visibility of photographers and photographic services within this institutional context. Although the library has had a photographic studio since 1910, this was staffed by librarians rather than professional photographers for the first fifty years of its existence. The article examines in turn three overlapping areas of activity: photographic reproduction for publications; the supply of facsimile copies to researchers; and the use of specialist imaging techniques. Although digital technologies have transformed imaging activities at the library, their core purpose remains – namely, to facilitate research with the library's collections.

Keywords: photography; library history; special collections; photomechanical processes

In his history of the John Rylands Library published in 1935, the Librarian Henry Guppy wrote:

> In the year 1910 the governors wisely decided to instal [sic] a photographic studio with a complete and up-to-date equipment of apparatus, and their action has been abundantly justified by the results already obtained. This new department is fraught with possibilities of world-wide benefit, for it has made it possible to render to scholars, both at home and abroad, most valuable assistance, by furnishing them with photographed facsimiles of pages from some of the rarer printed books and manuscripts. Again and again, in the case of requests for transcripts and collations of passages from some important text in the possession of the library, it has been found possible, at small cost, to provide a photograph or a rotograph of the passage required, which was at once more trustworthy and more acceptable than the best hand-made transcript could possibly be.[1]

Guppy's words are significant for outlining the primary function of photography at the Rylands during the twentieth century: to facilitate access to the collections for researchers who were unable to visit the library in person. Guppy was promoting photography primarily as a more efficient and 'trustworthy' alternative to copying texts by hand. His approach was shaped by an epistemology in which text was privileged over materiality, and an environment in which librarians mediated scholars' access to texts. Guppy did not explicitly reference another important function of certain photographic processes – to reveal information hidden from the human eye – even though the John Rylands Library had witnessed some tentative steps in this direction by the 1930s. Such purposes are now of central importance to the imaging practices discussed in the articles that follow, yet we argue that during the twentieth century, they were subordinate to the primary function of supplying images to researchers.

This article contextualises contemporary heritage imaging science at the John Rylands Library by exploring the twentieth-century development of its photographic services. We also explore the status and visibility of photographers and photographic services within the institutional context of the John Rylands Library. Although Guppy hailed the new 'department', our research reveals that during the first half-century of its existence, the service was operated by librarians rather than dedicated photographers. It was only in the 1960s, when technological developments demanded increased expertise from operators, that specialist photographers were employed in a discrete unit or department. This in turn resulted in the marginalisation of the photography department. Only in the last twenty years has the status of photographers risen, concomitant with the digital shift, as their expertise and input into research projects have come to be valued alongside those of researchers, librarians and archivists.

This paper is a work of 'archival history', interrogating the contexts of archives as well as their contents.[2] It is concerned with the relational meanings shaped by archival processes of creation, accumulation and preservation. The institutional archive of the John Rylands Library shapes the questions we are asking as well as supplying information and evidence. Records of photography and imaging are incomplete and those that survive are dispersed throughout the archive. These archival absences and dispersions mirror the liminal status of the photography department within the institution during the twentieth century. Interrogating the records that survive and the gaps in between, we focus on three areas of activity: the library's investment in the spaces and equipment required to produce images; the core purposes of collections-based photography; and the people who did the work. The patchy archival legacy makes it difficult to counter the historical invisibility of imaging specialists but does provide important clues to the history of imaging at the Rylands. This article, authored by professional photographers as well as archivists, re-inscribes photography back into the history of the Rylands.

We contend that advances in technology and practice during the twentieth century were driven primarily by the evolving requirements of researchers, rather than by technological innovation per se. Moreover, we aim to avoid the positivist trap of focusing on technical developments over the last century, culminating in today's 'state-of-the-art' facilities and processes. The history of heritage photography at the John Rylands Library is *not* a story of continual improvement; rather, it illuminates some of the persistent tensions and challenges that inhere in managing a research library with significant heritage collections, raising questions that continue to shape heritage imaging practice. By focusing on the 'why' rather than the 'how' and 'what', we hope to draw out parallels and continuities between earlier imaging activities at the library and contemporary practices.[3]

This article focuses on analogue photography of the library's special collections.[4] It considers in turn three overlapping areas of activity that have been paramount in the development of imaging at the library: photographic reproduction for publications, the supply of facsimile copies to researchers unable to consult materials in person and the use of specialist imaging techniques to enhance the visual experience and reveal information invisible to the human eye. It is important first to explore the antecedents of the photographic service, however, and to address the question of why a photographic studio did not feature in the initial plans for the Rylands Library.

Pre-history of the Imaging Service

Photography of the John Rylands Library's collections began before the library had even opened. The founder, Enriqueta Rylands, was a keen photographer, and one of the rooms at her residence – Longford Hall in Stretford – was set up as a 'photographic room', equipped with a full set of darkroom apparatus.[5] When Enriqueta Rylands purchased the world-famous Althorp Library in 1892, photographs were taken of highlights of the collections, including the 'Mentz Psalter', a Caxton indulgence, and a display of fifteen early blockbooks (Figure 1).[6]

The twenty-sixth Earl of Crawford, whose manuscript collection was acquired by Enriqueta Rylands in 1901, was another keen photographer. Glass-plate negatives of manuscripts and jewelled bindings had been created or commissioned by Crawford and by scholars, sometimes in connection with cataloguing projects that he sponsored.[7] Whether supporting research through the provision of clear and authentic reproductions, or promoting 'treasures' to a wider audience, the quality of images was a key concern. When Enriqueta Rylands purchased Crawford's collection, she was keen to acquire any photographs of the manuscripts; she recognised their importance and asserted her rights to them. She advised Alexander Railton of Sotheran's – her agent in the purchase – that 'it would be very objectionable when I have the collection if any firm could publish either photographs or catalogues or part of such without my knowledge or sanction'.[8] The value of these photographs would be realised in the completion and publication of the catalogues under Guppy's supervision.

In 1905, Enriqueta Rylands approved Guppy's proposal to publish a series of photographic facsimiles of some of the 'unique and rarer books' in the library.[9] These publications asserted the prestige of the library, facilitated book-historical

Figure 1 Photograph of blockbooks in the Althorp Library, taken for publication in *Black and White*, August 1892 (ref. VPH.325).

research and aided the preservation of unique items by reducing handling of the originals. Printing facsimiles was not a new endeavour, but Guppy wanted to take advantage of developing technologies to produce high-quality reproductions at an affordable price.[9] While the first facsimiles were produced, like the Demotic catalogue, by the Clarendon Press, Guppy also developed local collaborations, especially with Charles Gamble (1867–1951), the head of the photographic and printing department at the Municipal School of Technology.[10] This early cooperation between the John Rylands Library and what eventually became the University of Manchester presaged more recent scientific collaborations discussed in this issue. Publishing facsimiles was only appropriate for the rarest and most important objects, however; most photography responded to the needs of individual researchers, sometimes with very specialist interests.

In these circumstances, it is perhaps surprising that a photographic studio did not feature in the initial plans for the library, but photography was an expensive and specialised activity, and Guppy had more urgent priorities for the new institution: he needed to attract scholars to Manchester, assimilate and catalogue the vast and

Figure 2 Photography in the 1910 studio, *c.* 1954. On the right is Frank Taylor (1910–2000), Keeper of Manuscripts. The manuscript being photographed is Latin MS 104, a tenth-century copy of Smaragdus's Commentary on the Rule of St Benedict.

rapidly expanding collections and promote the collections to the wider public through exhibitions. In 1908, however, the library's finances were vastly augmented by a legacy of £200,000 from its founder. Two years later, the library established a photographic studio in a room originally intended as the kitchen to serve the refreshment room above (Figure 2). The studio remained there for half a century, until a major extension funded by the Wolfson family was constructed at the rear of the library in the early 1960s.[11] The remainder of this article will discuss how the photography service adapted to technological developments and the evolving needs of researchers in the twentieth century.

Photography for Publication

Throughout the library's history, an important aspect of its photographic activities has been the illustration of publications, from scholarly catalogues, monographs and facsimiles to exhibition catalogues and even postcards. The preparation of facsimiles and catalogues is well documented, especially in comparison to the daily work of the photography studio. One of the most significant early publications was Francis Llewellyn Griffith's catalogue of Demotic papyri in three large volumes, the first of which was an 'Atlas of facsimiles' containing 85 photographic plates.[12] The catalogue was commissioned by Lord Crawford but was not completed until 1909, after the manuscripts had been transferred to the John Rylands Library. Specialist techniques were required to capture and reproduce legible images of these papyri, since much of the writing is badly faded and some manuscripts are exceptionally large. The 'Great Peteesi Papyrus' (Demotic 9), over 4 metres long, still poses challenges to photographers today.[13] When Griffith first suggested an extensive suite of reproductions, in 1899, Crawford responded that they should only photograph a selection.[14] Griffith nevertheless argued that photography and hand-tracing were 'the only methods of any value' because there was no suitable type for printing demotic texts. He used photographs for the bulk of his transcriptions and analysis, though he still required sight of the originals, since 'the best papyri are too dark to depend on photographs alone for reading them'.[15]

The catalogue was to be printed by Oxford's Clarendon Press, led by Horace Hart, who had initiated significant developments at the press, including printing by lithographic and collotype processes; both were critical to the success of the Demotic catalogue.[16] Collotype was a process that created clear and precise reproductions from photographic negatives, without the pixelation inherent in the halftone process (Figure 3). Collotype printing was effectively a handicraft, however; the best facsimiles required time, skill and high-quality negatives. Hart sent a photographer to Crawford's Wigan residence, Haigh Hall, to take 'specimen negatives' for Griffith to review; although most were 'remarkably good', many were 'imperfect'. Griffith therefore suggested that the photographer should develop the negatives at Haigh Hall so they could be retaken on the spot if necessary.[17]

This level of specialist imaging and reproduction was expensive. Crawford met the costs of the photography at Haigh, but Enriqueta Rylands paid for Griffith's time, for the bulk of the photography, the making of plates, the printing, proofing and correcting, to the tune of nearly £1,000. Sadly, she died before the project was completed. Hart specified a binding 'worthy' of publishing them 'as a memorial of Mrs. Rylands', the cost of which was met from her estate.[18] Although the library continued to issue photographically illustrated catalogues, none were as comprehensive as those funded by the founder herself. Such technical and scholarly expertise requires substantial investment; for much of the history of the John Rylands Library, particularly as an independent institution, there were severe financial constraints which limited the capacity and capabilities of the imaging service. Many major projects to catalogue and digitise collections at the Rylands have only been made possible through funding from research grants, commercial partnerships and philanthropy. Although the internet has made it cheaper and easier to disseminate high-quality images – for example, through Manchester Digital Collections (MDC) – the resources required to create and curate them still constrain activity.[19]

'FRAUGHT WITH POSSIBILITIES OF WORLD-WIDE BENEFIT' 21

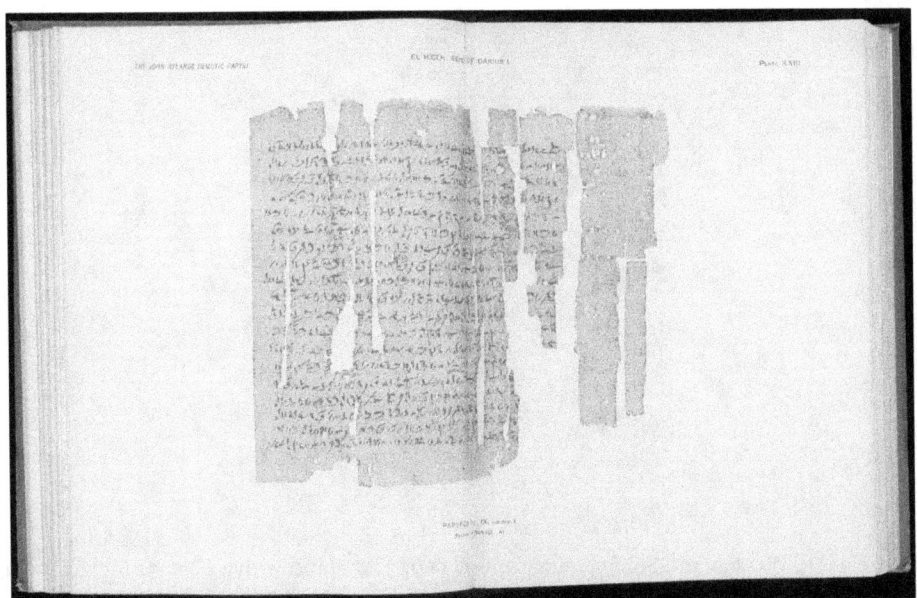

Figure 3 Plate XXIII in Griffith's *Catalogue of the Demotic Papyri in the John Rylands Library*: the first column of Demotic 9.

Photography for Access

Prior to 1910, researchers wishing to obtain photographs of items in the collections were advised to employ the services of external photographers. For example, in January 1909 the governors permitted the independent scholar John Wickham Legg to obtain 'rotary bromide facsimiles' of the illuminated Sarum missal manuscript and a printed missal of 1480; the photographer was advised to bring his own equipment.[20] We can assume that the use of external photographers presented practical challenges to the workings of the library and gave rise to questions over the ownership of negatives and of copyright in the images. Consequently, in 1909, the library developed plans to establish its own photographic studio. W. Watson & Sons Ltd of High Holborn, London, manufacturers of photographic apparatus, microscopes and optical instruments, supplied a 12 × 10-inch 'Premier' camera, 12-inch Holostigmat lens, right-angle reflecting prism and copying stand; a half-plate 'Alpha' camera and 7-inch Holostigmat lens; a half-plate enlarging lantern and pair of arc lamps; and other associated equipment. The equipment was expensive, costing £120, but it was specifically designed to efficiently reproduce books and manuscripts (Figure 4).[21]

In successive annual reports, Guppy claimed that the decision to create a photographic studio was 'abundantly justified' by the service it rendered to researchers.[22] The library still continued to rely on external specialists, however. When Oxford

Figure 4 W. Watson and Sons photograph of a copying stand with a 12 × 10-inch camera, 1909. JRL Archive, JRL/4/1/1/1909/Watson.

University Press required photographs for M. R. James's *Catalogue of Latin Manuscripts* in 1912, they sent their own operator to Manchester.[23] The library also employed local firms, such as J. T. Chapman Ltd and Flatters & Garnett, to develop negatives and to make prints and lantern slides.[24]

It is unclear how the studio operated in its early years. There was no photographic 'department', and the first designated 'photographer' was not employed until 1961.[25] Consequently, it is difficult to identify the individuals involved in producing images of the collections. The first reference to library staff involved in operating the studio occurs in the librarian's annual report for 1922, where it is stated that '[Thomas] Murgatroyd, one of the assistant-librarians, who took special training at the College of Technology, is in charge of the studio'.[26] In 1933, when Hyder Edwards Rollins of Harvard was supplied with rotographs of the Rylands copy of *England's Helicon* (London, 1600) in connection with his forthcoming edition of the work, he declared that they were 'so much more legible than those I received from the British Museum that I shall ultimately use them for printers' "copy" if you have no objection'.[27] Guppy ensured that this 'pleasing tribute to the work of Mr [Ronald] Hall [assistant librarian] and his assistant Mr [Lewis] Whittaker in the photographic studio' was brought to the attention of the library's governors.[28] By 1941, Moses Tyson, then Keeper of Western Manuscripts at the library, recorded that the 'excellent work' of Hall and Whittaker had 'resulted in scholars all over the world being supplied with photographs and photostats of rare books and manuscripts'.[29] The staff members involved in imaging were not photographers but librarians; they may have been given special training, but the evidence of surviving glass-plate negatives suggests that not all the photographs they took were of professional quality.

The development of specialist imaging techniques at the Rylands will be discussed later, but there are countless examples of the benefits that researchers derived from standard reproduction techniques. In 1934, Eugène Vinaver (1899–1979), Professor of French Language and Literature at the University of Manchester, negotiated the temporary deposit at the library of the newly discovered Winchester manuscript of Malory's *Morte d'Arthur* so that he could undertake a detailed textual comparison with the Caxton edition of 1485, of which the library held one of only two surviving copies.[30] The loan enabled Vinaver to make a side-by-side comparison of the two texts in Manchester, but he also requested rotographs of the manuscript, both to enable him to continue his studies while holidaying in France and so that the Clarendon Press could trial setting the type directly from the images.[31] Vinaver apologised to Moses Tyson: 'I am very sorry to give you all this trouble, but you know how much depends on this wretched MS.'[32] J. Pierpont Morgan also agreed to provide rotographs of his copy of the Caxton edition, enabling Vinaver to identify variant readings between the Manchester and New York copies.[33] Such comparative analysis of dispersed materials, enabled by photography, had a major impact on the development of textual scholarship. The library supported this type of research by improving and expanding its photographic facilities.

The 1910 studio proved inadequate to meet the increased demand for photographic reproductions in the postwar era. A new photography suite was therefore established in the 1960s on the ground floor of the Wolfson building, comprising a microfilm room, two copying rooms and a darkroom for processing and developing films and prints (Figure 5). These facilities reflected the priorities of the period, when microform was the most cost-effective means of making copies of works available beyond the physical library.[34] During the 1980s and early 1990s, the library entered into numerous partnerships with commercial organisations, such as Research Publications and Adam Matthew, to microfilm unique materials such as the Thrale-Piozzi Manuscripts, the Elizabeth Gaskell Manuscript Collection and several women's suffrage archives. In addition, the library contributed many thousands of printed books and pamphlets to collaborative microfilming initiatives, such as Research Publications' *Incunabula* project.[35] The large-scale image capture required for these projects was largely undertaken by external staff, both on- and off-site.

On-site processing facilities meant that the library was no longer reliant on external agencies to process its own microfilms and black and white prints, although colour printing continued to be outsourced to specialist laboratories such as Colour 061 in Salford.[36] Even with this large and well-planned photography suite, however, the John Rylands Library did not have the equipment or expertise required for specialist imaging, with one or two exceptions, discussed below.[37]

As noted above, the increasing technical complexity of image capture and processing led to the appointment of dedicated staff in the 1960s. Librarians were no

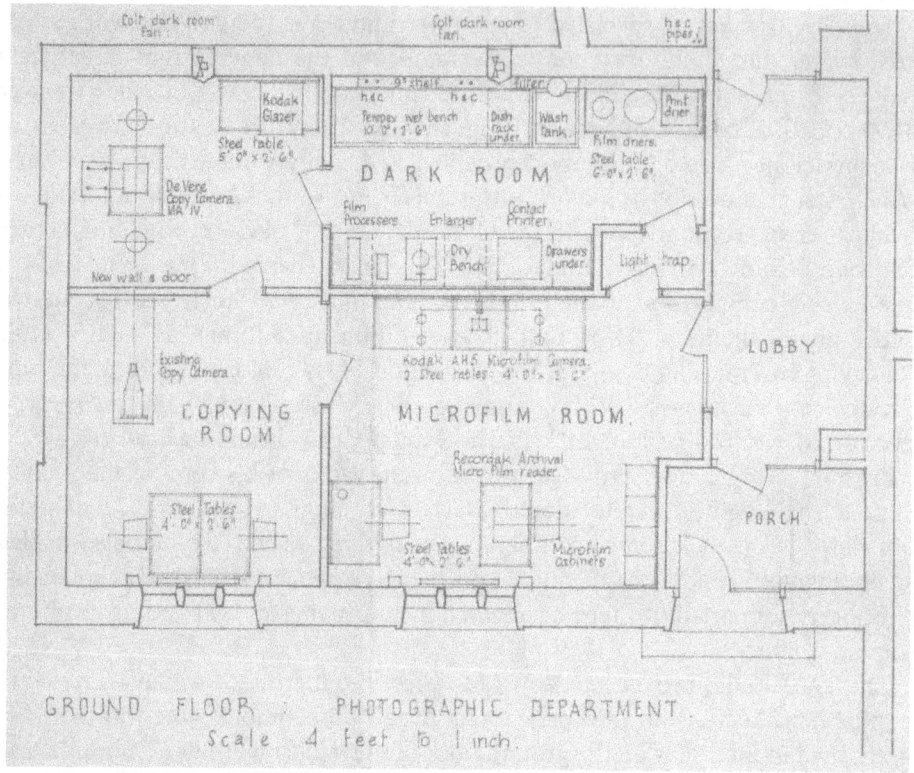

Figure 5 Francis Jones and Sons, Plan for the John Rylands Library Photographic Department, 1963. JRL Archive, JRL/5/4/4/27.

longer directly involved in the operation of the photography department, and the location of the studio at the rear of the building fostered a sense of marginalisation and separation from the general workings of the library. One of the authors remembers entering the studio with some trepidation in the early 1990s: one was not invited to pry too closely into the mysteries of what took place there, while the heady aroma of chemicals did not encourage one to linger. Moreover, library managers and curators regarded photographers as technical staff who could not be expected to show – and, indeed, were discouraged from showing – any curiosity or interest in the intellectual content of the material on which they worked; there was no knowledge exchange between the photographers and other library staff.

Specialist Imaging and New Technologies

Specialist photography responds both to the qualities of the materials being imaged and to the changing questions posed by researchers. The combination of material constraints and research curiosity is exemplified in the unique Mappemonde (world map) of 1546, ascribed to the distinguished French cartographer Pierre Desceliers and purchased by the twenty-fifth Earl of Crawford in 1874. The imaging history of this map began in August 1877, when Crawford gave permission to R. H. Major of the British Museum's Map Department to photograph it.[38] In 1895, in preparation for the publication of a facsimile edition of this and two other maps by Desceliers then in the British Museum, the twenty-sixth Earl of Crawford commissioned the Autotype Company in London to make large glass-plate photographs of all three maps.[39] Producing high-quality images of such large maps required the maps to be photographed in multiple sections. All of these – forty-nine in total – were delivered to Enriqueta Rylands in October 1901 and transferred to the library in January 1905.[40] The printed reproductions were difficult to access, however, and although they were remarkably detailed, they did not meet the needs of researchers.

Advanced imaging technologies can enhance the visual experience, sometimes revealing what is hidden from the human eye, such as the undertext of a palimpsest. The damaging chemical reagents used by nineteenth-century researchers to reveal – albeit fleetingly – these hidden texts were replaced in the twentieth century by ultra-violet (UV), analogue X-ray and infrared imaging tools, using light frequencies outside the range of human vision.[41] In 1950, Cottie Burland of the Department of Ethnography at the British Museum contacted the John Rylands Library requesting photographs of the Mappemonde, and of the Canadian portion of the map in particular, 'because it seems a most promising object of study in connection with early American history, particularly of the first contacts between Indians and the white men'.[42] He suggested that satisfactory results might be obtained by photographing some of the darkened and discoloured areas under ultraviolet light. Edward Robertson, the new Rylands Librarian, explained that the library had no facilities for UV photography, and subsequent enquiries among Manchester photographers revealed that none possessed the requisite equipment. Burland therefore arranged for a London firm of technical photographers, Messrs R. B. Fleming & Co. Ltd, to undertake the photography at the library.[43] In fact, Burland subsequently reported that while the map was examined under UV, the photographs were taken by infrared light, which 'resulted in a considerable improvement in the definition obtained'.[44] Indeed, a comparison of Figures 6 and 7 reveals a substantial increase in legibility.

Beta-radiography was another technology employed at the Rylands and other major libraries, most commonly used to reveal watermarks and the laid and chain lines that constitute the 'fingerprints' of specific paper stocks. The sheet of paper to be examined was positioned between a low-energy carbon-14 radiation source embedded in a polymer sheet, and an X-ray film on to which the image formed.[45] We do not know when this process was introduced at the Rylands, but it was used by Alan Crown in the 1980s to analyse the morphology of the paper used in Samaritan manuscripts held at the John Rylands Library, the Bodleian Library and the Bibliothèque Nationale in Paris.[46] Due to the long exposure times required and safety concerns surrounding radioactive materials, beta-radiography has been

Figure 6 Detail of autotype facsimile of the Mappemonde, in Coote, 1898.

Figure 7 Infrared photograph in C. A. Burland, 'A Map of Canada in 1546', 1951.

entirely superseded by advanced digital imaging technologies, as shown by the article by Stephen Mossman and Edward Potten in this issue.

Analogue methods such as UV lamps, microscopes and light sheets are still used by researchers and conservators, but have largely been superseded by advanced digital technologies.[47] These technologies have also enabled new modes of imaging heritage objects beyond the limitations of static, 2D photographs. Cuneiform tablets, in which the text is impressed in the clay, require a different approach to letters inked on a page, for example. When the library's large cuneiform collection was digitised in 2012 by the Cuneiform Digital Library Initiative, about a quarter were imaged with reflectance transformation imaging (RTI) with the assistance of a photo dome.[48] In this case, RTI represents the materiality and contours of 3D objects, imitating the 'real life' experience of viewing the original objects. The exciting possibilities of RTI used in conjunction with other imaging technologies are explored by Stefan Hanß et al. in this issue. While the library is now in the forefront of advanced digital imaging technologies, however, it is evident that for much of the twentieth century, the library's facilities were limited by a lack of resources and perhaps a lack of demand. Researchers who sought to reveal hidden texts or images through specialised imaging techniques were obliged to involve external agencies.

Conclusion

Heritage photography is folded into the fabric of the John Rylands Library. Enriqueta Rylands commissioned a series of photographs of portraits to aid in the design of the sculpted figures in the library, including the portrait of Shakespeare in her copy of the First Folio (Figure 8).[49] We now have the tools to create 3D laser-scanned models of the collections, and even of the entire building.[50] Alongside innovative technology comes a recognition of the importance of historic practices; old photographs of the collections are increasingly valued by researchers as traces of the history of the collections.[51]

In the last twenty years, thanks to the shift to digital technologies and advanced imaging techniques, and much greater academic interest in the special collections, both within the university and internationally, our imaging service has undergone a complete transformation. Professional photographers were recruited in the mid-2000s to support major research projects, the first of which involved the digitisation and cataloguing of the Rylands Cairo Genizah Collection. The team has grown to comprise an imaging manager, two senior photographers, a photographer and two photographic assistants, who collaborate with other teams across the library on initiatives such as Manchester Digital Collections. Another important development has been the concept of preservation photography. In the twentieth century, efficiency was privileged over preservation concerns: bulldog clips kept pages secure and bright 'mercury-lamps' reduced exposure times. Our practices now prioritise the physical care of the object, from conservation-grade supports and low UV levels to the production and preservation of high-resolution images fit for a wide variety of purposes.

Figure 8 Photograph of frontispiece of Enriqueta Rylands's Shakespeare First Folio, c. 1896. JRL Archive, JRL/5/2/3.

In late 2024, the library opened a brand-new Advanced Imaging Laboratory on the first floor of the Wolfson building. This massively expands our capabilities for digitisation and advanced imaging, and enables researchers to work alongside photographers in more comfortable and spacious surroundings. One of the first projects planned involves the conservation and advanced imaging of the large Mappemonde, encompassing 3D imaging (to capture the contours of the cockled parchment) as well as multispectral imaging (to reveal faded, obscured or damaged areas). While imaging necessarily takes place in a dedicated space, the department is now better integrated into the wider library and the research life of the university. This may in part be because the alchemy of analogue photography has been superseded by digital photography, which involves skillsets more closely aligned with those of information professionals and researchers. Digital photographic processes are also more accessible to researchers and others, who can view images as they are created, working side by side with photographers in ways that were impossible in the analogue era.

Imaging at the Rylands has altered in many ways since the first studio opened in 1910, but its core purpose remains – namely, to facilitate research with the library's collections. The ways in which 'images' (or visual data) can be recorded, analysed, shared and preserved are certainly different in a digital world. Nevertheless, while all the articles in this issue refer to the use of digital technologies, the skills and expertise of the *people* doing the work – photographers, analysts, scientists, researchers and curators – remain critical to success. Indeed, some of these projects can be recognised as developments of the types of activity, over the past century and more, described in this article: the digitisation of Gaster Amulets is part of a large cataloguing project; the analysis of the Trier psalter-hymnal puts a manuscript into its historical context; and the identification of watermarks uses technology to reveal what is hidden from the naked eye. At the same time, changing research questions and developing technologies have put the materiality and histories of text-bearing objects on an equal footing with the texts they present. The articles also speak to broader developments in heritage practice. Thus, Gurtek Singh and James Robinson, in their discussion of the digitisation of Sikh cultural and religious heritage, show that heritage imaging is not neutral but has social and ethical implications. We cannot predict how imaging at the Rylands will change over the next hundred years any more than Guppy could; yet this special issue convinces us that the future is assured, exciting and still 'fraught with possibilities'.

Notes

1. H. Guppy, *The John Rylands Library: 1899–1935: A Brief Record of its History, with Descriptions of the Building and its Contents* (Manchester: Manchester University Press, 1935), p. 23.
2. T. Cook, 'The Archive(s) Is a Foreign Country: Historians, Archivists, and the Changing Archival Landscape', *The American Archivist*, 74:2 (2011), 600–32.
3. James Freeman adopted this approach in his study of R. B. Haselden, who pioneered innovative techniques for studying manuscripts at the Huntington Library, California: 'Scientific Aids for the Study of Manuscripts: R. B. Haselden and the History of Heritage Science', *Heritage Science: Are Manuscript Studies Experiencing a 'Scientific Turn'?* virtual seminar, Association for Manuscripts and Archives in Research Collections, 28 April 2023, https://amarcsite.wordpress.com/heritage-science-2023/ [accessed 4 October 2024].
4. The no-less-important photography of the library's architectural and iconographical features lies outside the scope of this article.
5. For a detailed list, see *Catalogue of Excellent Modern Household Furniture [. . .] which will be Sold by Auction [. . .] on the Premises, Longford Hall, Stretford* (Manchester: Capes, Dunn, 1908), University of Manchester Library (hereafter 'UML'), Orford Papers, ORF/3/1/42, pp. 65–7.
6. The photographs were published in *Black and White: A Weekly Illustrated Record and Review*, 20 August 1892, pp. 210–12, and 27 August 1892, p. 257. A complete set of prints was recently discovered in the library (ref. VPH.325).
7. UML, John Rylands Library Archive (hereafter JRL Archive), JRL/6/1/6/1/1, letter from Crawford to Railton (31 July 1901). Edinburgh, National Library of Scotland

(hereafter NLS), Acc. 9769, Crawford Papers, Crawford Library Letters (hereafter CLL), vol. 53, f. 663, letter from J. P. Edmond to Crawford (20 September 1895).

8 JRL Archive, JRL/6/1/6/1/1, draft letter from Enriqueta Rylands to Railton (1 August 1901).

9 See Guppy's list of 'Facsimiles of Manuscripts', which range in date from the 1830s to the 1900s, in *Catalogue of an Exhibition of Illuminated Manuscripts, Principally Biblical and Liturgical* (Manchester: John Rylands Library, 1908), pp. 51–7.

10 JRL Archive, JRL/4/1/1/1906/Gamble, letter from Gamble to Guppy (13 June 1906). The preface to the first facsimile, *Propositio Johannis Russell printed by William Caxton* (1909), records that it 'has been prepared by a photographic method of line engraving, and printed typographically, in the Photography and Printing Crafts Department of the School of Technology'.

11 F. Taylor, 'The John Rylands Library, 1936–72', *Bulletin of the John Rylands Library*, 71:2 (1989), 64–5, https://doi.org/10.7227/BJRL.71.2.2.

12 F. Griffith, *Catalogue of the Demotic Papyri in the John Rylands Library, Manchester* (Manchester: Manchester University Press, 1909). The photography and reproduction of these plates is documented in: NLS, CLL, vols 72–8, 1899–1900; Enriqueta Rylands's papers: JRL Archive, JRL/6/1/6/2-3; Henry Guppy's papers: JRL Archive, JRL/6/4/4. Other catalogues with plates likely derived from Crawford photographs include M. R. James, *A Descriptive Catalogue of the Latin Manuscripts in the John Rylands Library* (Manchester: Manchester University Press, 1921); W. E. Crum, *Catalogue of the Coptic Manuscripts in the Collection of the John Rylands Library* (Manchester: Manchester University Press, 1909); A. S. Hunt, *Catalogue of the Greek and Latin Papyri in the John Rylands Library*, vol. 1 (Manchester: Manchester University Press, 1911).

13 J. Offord, 'The Great Peteesi Papyrus', *The American Antiquarian and Oriental Journal*, 35:3 (1913), 162; National Manuscripts Conservation Trust, 'Case Study: John Rylands Research Library – The Conservation of Demotic Papyrus 9', 2022, www.nmct.co.uk/case-studies/john-rylands-research-library [accessed 29 January 2025].

14 NLS, CLL, vol. 72, ff. 622 and 631, letter from Griffith to J. P. Edmond (8 May 1899), letter from Crawford to Edmond (9 May 1899).

15 NLS, CLL, vol. 72, f. 715, letter from Griffith to Crawford (31 May 1899).

16 R. M. Ritter, 'Horace Hart and the University Press, Oxford, 1883–1915', *Journal of the Printing Historical Society*, NS 7 (2004), 23–36.

17 NLS, CLL, vol. 78, October–December 1900, f. 840, letter from Griffith to J. P. Edmond (undated [*c.* 1 October 1900]).

18 JRL Archive, JRL/6/4/4, letter from Hart to Guppy (23 September 1908), sending estimates for cost of completion of Demotic and Coptic; JRL/1/3/1/1, Annual Report, 1908.

19 MDC is branded as a 'new resource for exploring high-quality images of cultural collections and research projects'; see University of Manchester Library website, 'Introducing Manchester Digital Collections', www.digitalcollections.manchester.ac.uk/about/ [accessed 27 September 2024].

20 Missal (Sarum), Latin MS 24, available online at www.digitalcollections.manchester.ac.uk/view/MS-LATIN-00024/1 [accessed 1 October 2024]; *Missale Messanense*

secundum consuetudinem Gallicorum (Messina) (Messina: Henricus Alding, 1480), 18620, available online at https://luna.manchester.ac.uk/luna/servlet/s/2yy9ll [accessed 1 October 2024]. JRL Archive, JRL/1/3/2/1, Council minute book, p. 284, 25 January 1909.

21 JRL Archive, JRL/4/1/1/1909/Watson, papers relating to estimate from Watson & Sons, June–September 1909. We are grateful to Prof. Stephen Milner for drawing this material to our attention. The House Committee approved the estimate on 14 June (JRL/1/3/5/1, minutes of House Committee). Watson camera equipment is described and illustrated on the Early Photography website, www.earlyphotography.co.uk/index.html [accessed 29 January 2025]. See also T. Richards, 'Histories of Heritage Imaging', University of Manchester digital repository, 2022, https://doi.org/10.48420/20073470.v1.

22 Including, in 1911, the passage he later reworked into his history of the library, quoted at the beginning of this article. JRL Archive, JRL/1/3/1/1, annual report of the Librarian for 1911, p. 4.

23 JRL Archive, JRL/4/1/1/1912/Hart, letter from Hart to Guppy (18 May 1912).

24 JRL Archive, JRL/4/1/1/1912/Chapman, letter from J. T. Chapman Ltd to [Oliver] Sutton, [assistant librarian] (22 October 1912); JRL/4/1/1/1908/Flatters, invoice from Flatters & Garnett (8 September 1908). See also the Science Museum Group Collection website, https://collection.sciencemuseumgroup.org.uk/people/ap14523/j-t-chapman-ltd and https://collection.sciencemuseumgroup.org.uk/people/cp3558/flatters-and-garnett-limited respectively [accessed 8 October 2024].

25 Edwin (or Edward) Beaumont Bathe was appointed as a part-time photographer from February 1961. He retired in 1973 at the age of 70. JRL Archive, JRL/3/2, Superannuation Fund contribution ledgers; JRL/3/6/2 records relating to salary reviews and contracts; 'Librarian's Report, 1973–74', p. 8, *Bulletin of the John Rylands Library*, 57:2 (1975), https://doi.org/10.7227/BJRL.57.2.11.

26 JRL Archive, JRL/1/3/1/2, annual report of the Librarian for 1922. Similarly, the British Museum Library did not employ dedicated professional staff until 1927. A. Esdaile, *The British Museum Library: A Short History and Survey* (London: George Allen & Unwin, 1946), p. 343.

27 JRL Archive, JRL/4/1/1/1933/Rollins, letter from Rollins to Guppy (11 February 1933). *England's Helicon, 1600, 1614*, ed. H. E. Rollins (Cambridge, MA: Harvard University Press, 1935). Rotography is a photographic technique whereby an image is projected on to a roll of sensitised paper, by means of a mirror and lens, to create a full-size reproduction in which positive and negative tones are reversed. See H. W. Ballou, 'Photography and the Library', *Library Trends*, 5:2 (1956), 265–93 (268–70), http://hdl.handle.net/2142/5689 [accessed 10 January 2025]; J. H. Teper, 'Popular 20th Century Office Reprographic Processes – A Guide to Identification and Preservation', *Collections*, 3:1 (2007), 17–18, https://doi.org/10.1177/155019060700300103.

28 JRL Archive, JRL/1/3/2/4, Council minute book, p. 360 (27 February 1933).

29 M. Tyson, 'The First Forty Years of the John Rylands Library', *Bulletin of the John Rylands Library*, 25:1 (1941), 5, https://doi.org/10.7227/BJRL.25.1.6.

30 E. Vinaver, 'Malory's Morte D'Arthur in the Light of a Recent Discovery', *Bulletin of the John Rylands Library*, 19:2 (1935), 438–57, https://doi.org/10.7227/BJRL.19.2.10.

See also E. Vinaver, *The Works of Sir Thomas Malory* (Oxford: Clarendon Press, 1947), which includes three images of the manuscript (now British Library Add. MS 59678).

31 Although this attempt was unsuccessful, the images were subsequently used to illustrate Vinaver's edition, *The Works of Sir Thomas Malory* (1947).

32 JRL Archive, JRL/4/1/1/1934/Vinaver, letters from Vinaver to Tyson (15 July and 11 October 1934).

33 JRL Archive, JRL/4/1/1/1934/Vinaver, letter from Vinaver to Guppy (25 June 1935).

34 For an overview of the development of microforms, see Ballou, 'Photography and the Library'; Teper, 'Popular 20th Century Office Reprographic Processes'.

35 P. McNiven, 'The John Rylands Library 1972–2000', *Bulletin of the John Rylands Library*, 82:2–3 (2000), 31, https://doi.org/10.7227/BJRL.82.2-3.1. McNiven notes that the Rylands was previously reluctant to contribute to collaborative microform projects, fearing that they would discourage researchers from visiting the library.

36 Personal recollection of J. Hodgson.

37 The University Library (styled the Main Library after the merger in 1972) also housed a small photographic unit. In 2007, photography services at the John Rylands Library reverted to the 1910 studio, following the repurposing of the Wolfson building.

38 C. H. Coote, *Bibliotheca Lindesiana: Collations and Notes No. 4, Autotype Facsimiles of Three Mappemondes* (privately printed, 1898), p. 12.

39 Crawford asked the Autotype Co. to supply 100 collotype copies of each of the fifteen plates of the Desceliers Mappemonde, as per their estimate of £105. NLS, CLL, vol. 54, October–December 1895, f. 962, copy letter from J. P. Edmond to the Autotype Co. (30 December 1895). The plates were published (along with facsimiles of two maps in the British Museum) in Coote, *Bibliotheca Lindesiana*, p. 12. On the Autotype Co., see J. Moore, *Celebration of Innovation: A History of Autotype 1868-2005* (Wantage: Autotype International, 2005). The Rylands Mappemonde is now French MS 1*; digital images of the map are available online at https://luna.manchester.ac.uk/luna/servlet/s/sk4h4x [accessed 8 October 2024].

40 They remain in the library today, their storage and management providing an equal but different challenge to that of the map.

41 On chemical reagents, see F. Albrecht, 'Between Boon and Bane: The Use of Chemical Reagents in Palimpsest Research in the Nineteenth Century', in *Care and Conservation of Manuscripts 13: Proceedings of the Thirteenth International Seminar Held at the University of Copenhagen, 13th–15th April 2011*, ed. M. J. Driscoll (Copenhagen: Museum Tusculanum Press, 2012), pp. 147–65.

42 JRL Archive, JRL/4/1/1/1950/Burland, letter from C. A. Burland to the John Rylands Library (6 July 1950). On Burland, see E. Carmichael, 'Obituary: C. A. Burland', *RAIN*, 56 (1983), 15–16, www.jstor.org/stable/3033432 [accessed 24 January 2025].

43 JRL Archive, JRL/4/1/1/1950/Burland, copy letters from the library to Burland (7 July and 3 October 1950); letters from Burland to the library (27 September and 6 November 1950).

44 C. A. Burland, 'A Note on the Desceliers' Mappemonde of 1546 in the John Rylands Library', *Bulletin of the John Rylands Library*, 33:2 (1951), 237–41 (238), https://doi.org/10.7227/BJRL.33.2.4. The photographs were reproduced (in monochrome) in C. A. Burland, 'A Map of Canada in 1546', *Geographical Magazine*, 24 (1951), 103–10.

45 N. E. Ash, 'Recording Watermarks by Beta-Radiography and Other Means', *Book and Paper Group Post Prints* (Washington, DC: American Institute for Conservation, 1982), https://cool.culturalheritage.org/coolaic/sg/bpg/annual/v01/bp01-02.html [accessed 10 October 2024].

46 A. D. Crown, 'The Morphology of Paper in Samaritan Manuscripts: A Diachronic Profile', *Bulletin of the John Rylands Library*, 71:1 (1989), 71–94, https://doi.org/10.7227/BJRL.71.1.5. The radioactive source was removed from the library in *c.* 2001.

47 Because these practices do not produce physical image objects, they are poorly represented in the archival record.

48 See E. Gow, 'Digitisation of Ancient Cuneiform Texts', *Rylands Blog* (12 November 2012), https://rylandscollections.com/2012/11/13/digitisation-of-ancient-cuneiform-texts/ [accessed 10 October 2024].

49 JRL Archive, JRL/5/2/3, photographs and correspondence, 1893–7. The visible fold line corresponds to the copy of the First Folio sold in the 1930s and now at the Lilly Library, Indiana University. It should also be noted that Rylands commissioned the architectural photographers Bedford Lemere to take a series of photographs of the newly opened library in 1900.

50 T. Richards, 'From 2D to 3D – Photogrammetry Part 2: Investigating 3D Imaging of our Special Collections', *Rylands Blog* (16 December 2020), https://rylandscollections.com/2020/12/16/from-2d-to-3d-photogrammetry-part-2/ [accessed 11 October 2024].

51 A project has begun to conserve and digitise the library's glass-plate negatives. See project blog posts: https://medium.com/special-collections/histories-of-heritage-imaging-at-the-john-rylands-library-part-1-3974c7a3db01, https://medium.com/special-collections/histories-of-heritage-imaging-at-the-john-rylands-library-part-2-514907958954 and https://medium.com/special-collections/histories-of-heritage-imaging-at-the-john-rylands-library-part-3-26e70cf449ee [accessed 29 January 2025].

Imaging Heritage Featherwork: A New Methodology for the Study of Feather Artefacts

STEFAN HANß, UNIVERSITY OF MANCHESTER
JAMES ROBINSON, UNIVERSITY OF MANCHESTER
TONY RICHARDS, UNIVERSITY OF MANCHESTER

Abstract

The article discusses the use of imaging methodology in research on featherwork heritage. The project team of historians, imaging scientists and photographers, conservators and curators was the first to systematically develop an advanced imaging approach to the study of historic featherwork. The object of analysis was a highly elaborate feather fan produced either in the Low Countries or in Dutch colonial Brazil, likely around 1665, that is today held in the Fitzwilliam Museum, University of Cambridge. Building on existing scholarship by biologists, conservators and ornithologists, the article discusses the extent to which imaging approaches help contextualise discussions about photosensitivity, pigmentation and the fluorescence of feathers as well as species identification. Imaging also allows for new insights into crafts knowledge, intricate manufacturing skills as well as invisible artisanal interventions, such as the use of adhesives. Moreover, imaging generates new opportunities for the conservation and display of featherwork and provides momentum for novel collaborations to be built with Indigenous descendant communities and feather practitioners today.

Keywords: feathers; imaging; heritage science; conservation; Indigeneity

Imaging beyond the Written or Illuminated Page

Imaging has revolutionised heritage science in recent years. Imaging analysis is a well-established element in the study of historical manuscripts and prints as such non-invasive methodological approaches provide step-change insights into the recovery of lost or palimpsest texts, pigment and ink analysis as well as watermark, paper and binding analysis, among others. In more recent years, however, advanced imaging techniques have increasingly been used to unlock heritage artefacts beyond the written page. Paintings and drawings in museums across the world have been studied with multispectral imaging to uncover hidden or invisible underdrawings.[1] Polychrome statues too have been subject to new imaging approaches, resulting in major methodological advancements in the digital mapping of luminescent materials like consolidants, organic binders, varnishes and pigments, such as those used in Hellenistic terracotta statuettes from Italy and Turkey.[2] 3D imaging has also provoked wider reflections on the ways that visual representation impact archaeological research questions and methodologies, prompting archaeologists to search for new ways to display research findings.[3] Moreover, multispectral imaging has been applied to map the photoluminescence and reflectance qualities of textiles from late antiquity, resulting in new insights into the use of red and blue dyestuffs and more targeted analysis of these pigments.[4] Multispectral imaging has also proved helpful in conservation works of medieval glass wall mosaics.[5] Recently, researchers have even studied natural museum specimens like lepidoptera collections with multispectral imaging, emphasising the potential of enhanced imaging to shift taxonomic research.[6] As this incomplete list of examples illustrates, advanced imaging provides major opportunities for innovation and potential for methodological advancements in wider heritage science research, beyond the analysis of the written and illuminated page of manuscripts or prints.

The Imaging Laboratory Team of the John Rylands Research Institute and Library, University of Manchester, is spearheading some of these heritage science interventions – for instance, when experimenting with different kinds of heritage artefacts and materials or contributing to broader comparative research on multispectral imaging of cultural heritage across institutions worldwide to establish sector-wide best practice.[7] This article illustrates an example of collaboration across collections, disciplines and institutions, as well as the insights that Rylands imaging science expertise can bring to research on heritage artefacts beyond library contexts. Building on existing scholarship by biologists, conservators and ornithologists who evidenced spectral curve responses of feathers to different wavelengths, the project's principal investigator (PI) at the University of Manchester partnered with the John Rylands Research Institute and Library Imaging Lab to connect UK-wide leading imaging technology with heritage collections across the country to advance the analysis of historic featherwork.

The project team of historians, imaging scientists and photographers, conservators and curators was the first to systematically develop an advanced imaging approach to the study of historical featherwork (Figure 1a). The object of analysis was the Messel Standing Feather Fan from the Fitzwilliam Museum, University of Cambridge (M.358-1985): a highly elaborate feather fan produced either in the Low Countries or in Dutch colonial Brazil, likely around 1665, named after its former owner Leonard Messel (1872–1953). The project principal investigator (PI) has worked on featherwork from across the early modern globe, circa 1400–1850, for years and had studied the Messel Standing Feather Fan on various occasions since 2016, yet collaboration materialised with the submission of a formal request for technical examination and scientific analysis to the Fitzwilliam Museum in 2019. The support, enthusiasm and open-mindedness of the curators, conservators and heritage scientists of the museum was crucial in implementing this project, as was the expertise, energy and creativity of the John Rylands Research Institute and Library Imaging Lab team. The project resulted in unprecedented insights into the cross-cultural world of early modern featherwork, published in *Current Anthropology* in 2024.[8] The present article complements the discussion of the fan as a site of

 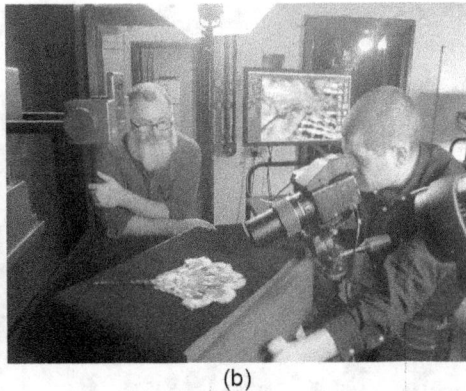

Figure 1 (a) The Messel Standing Feather Fan at the Imaging Lab at the John Rylands Research Institute and Library, University of Manchester, with Jo Dillon, James Robinson, Tony Richards, Stefan Hanß and Gwen Riley Jones. (b) James Robinson and Tony Richards conducting macrophotography.

innovative cultural crossings and material experiments in the age of early modern globalisation, consumerism and colonialism in that publication by highlighting the project's methodological intervention in the wider field of imaging heritage science.

Imaging Techniques and System Specifications

Imaging approaches, we argue, provide new insights into heritage featherwork that hold transformative opportunities for the study, preservation and display of such artefacts. Advanced imaging reveals novel understanding of materials, crafts and artisans' decision-making, as well as the artefacts' history and object biographies. This approach uncovers largely oral and often partially lost crafts knowledge with wider cultural significance, as well as knowledge and information invisible to the naked human eye. Moreover, advanced imaging approaches recover information of vital importance to the future preservation of such critically endangered heritage. These research insights, then, hold wider repercussions for the display of featherwork in museums, exhibitions and online collection servers worldwide.

This study applied different imaging techniques. High-resolution macrophotography resulted in images of unprecedented depth and detail, and it was used alongside focus stacking (Figure 1b). This digital image-processing technique combines multiple images of different focal distances into one composite image. The resulting image flattens depth of field to ensure that each area is in focus. We also used transmitted light imaging – that is, a flat copy-imaging technique in which a Phase One XF camera (IQ4 digital back, 120-mm Schneider lens) was mounted above the Messel Standing Feather Fan while lighting was installed either at 45° angles or transmitted through the object.

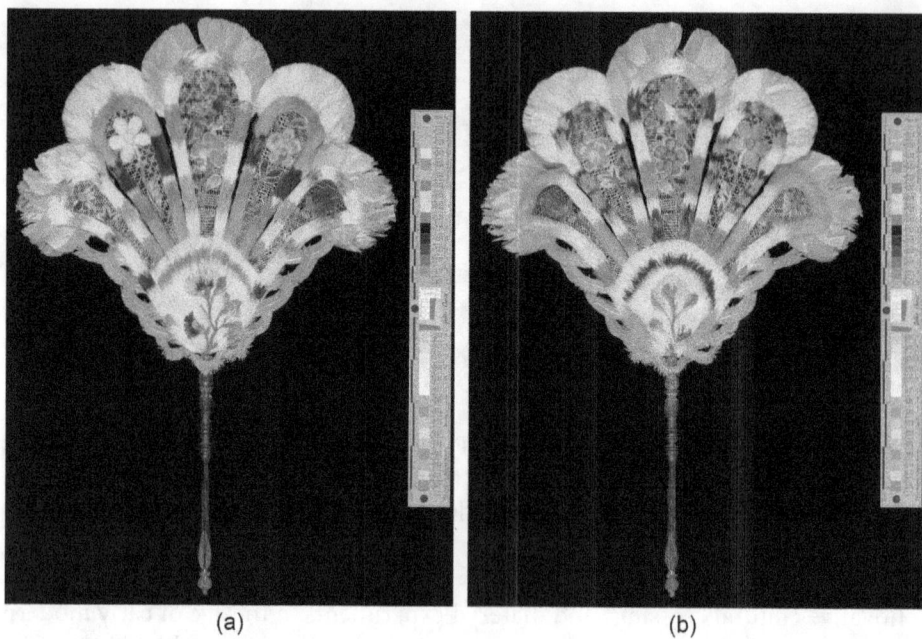

Figure 2 Messel Standing Feather Fan, both sides. 34 × 23 cm. Fitzwilliam Museum Cambridge, M.358-1985. Images © The John Rylands Research Institute and Library, University of Manchester, reproduced by permission of the Fitzwilliam Museum, University of Cambridge.

In addition, multispectral imaging (MSI) analysis was performed using a Phase One IQ260 digital system (achromatic, 60-megapixel, 16-bit monochrome digital back with 8,964 × 6,716-pixel charge-coupled device array at 6.0-μm pixel size; Phase One iXr camera body and 120-mm Schneider lens) to reveal infrared and ultraviolet wavelengths (365–940 nm). The system contained an Equipoise Imaging filter wheel with LP400 (long pass violet), LP515 (long pass green) and LP590 (long pass red) filters to better capture iridescence and reflectance responses. Equipoise Imaging light panels provided narrowband light-emitting diode illumination of the object in sixteen narrow wavelength bands. Such analysis resulted in twenty-two images of sixteen different wavelengths and five additional images with long pass violet, green and red filters for indicating fluorescence. MSI analysis was conducted on both sides of the Messel Standing Feather Fan (for a full view of the fan, see Figure 2a–b), and for eight additional detailed views.

To create 3D views of the fan, we processed photographs of the artefact with different lighting options and from different angles using reflectance transformation imaging (RTI), also known as polynomial texture mapping (PTM), 'a computational photographic method that captures a subject's surface shape and color and enables the interactive re-lighting of the subject from any direction' and that 'also permits the mathematical enhancement of the subject's surface shape and color attributes'. The viewer software has been developed by CHI Cultural Heritage Imaging and is available open access.[9]

To make informed decisions regarding project design and imaging analysis, we also used a Dino-Lite USB digital microscope AM7013MZT with polariser, adjustable zooming and lighting options. The PI was among the first historians to use digital microscopes in the study of early modern material culture, namely feathers and textiles, due to the potential of microscopic analysis to provide unique material insights into the production, use and experience of artefacts and to create more inclusive material culture studies. The digital microscope 'is a historian's indispensable aid for examining the material properties of feathers as well as the craftsmanship of featherworkers, and thus for studying the matter and making of featherwork and how it was experienced on a sensory level'.[10] In the present study, digital microscopy facilitated informed decision-making about where to proceed with high-resolution macrophotography and MSI.

The study was enhanced by in-depth archival research and the study of a wide range of featherwork preserved in museum collections worldwide, which provided indispensable contextualisation to situate specific material practices in particular cultural, social and historical milieus. This research has been informed by conversations with practitioners as the material experience of feather artists has been essential for the interpretation of scientific data.[11]

Materials

MSI analysis of historical featherwork provides unique insights into the pigmentation and fluorescence of processed feathers. Such insights have repercussions for future improvements in species and feather identification. Although we are yet to fully understand the ways that feather pigmentation and colour schemes respond to different wavelengths, the research of biologists, conservators and ornithologists has shown that feather pigments have bespoke spectral responses. Such research stems from the works of evolutionary biologists who examined correlations between the visual ecology of feathers and the distribution of retinal cone photoreceptors in different bird species. The different design of the eyes of various bird species, such research shows, impacts the ways feathers are seen by birds as well as the wider development of feather pigmentation across species. Since the distribution of cone photoreceptors echoes foraging strategies and habitats of different avian species, biologists have become interested in the likely link between the evolution and behaviour of birds and the visual ecology of feathers, like colour pigmentation schemes or iridescence.[12] For instance, biologists have sought to examine whether fluorescence is a mere by-product of the pigment structure of feathers or whether it holds wider significance in avian behaviour, with experiments confirming that fluorescence significantly impacts birds' sexual preference.[13] Species-specific 'spectral tuning' analysis has been conducted for different birds since the beginnings of such research, yet most important for our contexts is the fact that such evolutionary biological research has resulted in graphs showcasing the spectral curve responses of different species' feathers of various pigmentation. Feathers have bespoke spectral sensitivities, and their pigmentation translates into specific reflectance spectra.[14]

Biologists' research on different feather colourants and their respective degrees of photosensitivity has significant repercussions for the work of conservators. Colourants have a bespoke response to light exposure, so conservators have drawn on such biological research to develop better lighting policies for the museum display of featherwork.[15] This has resulted in a call for advanced colour measurements and pigmentation analysis of feathers, which has in turn confirmed that feather pigments correlate with specific spectral responses. In pivotal publications, conservator Ellen Pearlstein has argued that ultraviolet illumination facilitates both the identification of feather species and light-induced degradation.[16] Likewise, infrared imaging has proven a methodologically innovative tool to reveal underdrawings in Mexican feather mosaics.[17] Such research has informed our advanced imaging analysis of the Messel Standing Feather Fan.

MSI analysis of the fan confirms that feathers have bespoke spectral curve responses, and that these can potentially be linked to pigment and species identification (Figure 3a). If contextualised with published data, the microphotographic analysis of the fan and ornithological specimens, MSI provides rare insights into surface structure, colour and pigmentation specificities, iridescence patterns and spectral sensitivities that enhance avian identification. The Messel Feather Fan artist predominantly used American birds to craft the fan, and here in particular feathers of birds whose habitats map on to the territories of early modern Dutch colonies in Brazil and the Guianas. We identified cobalt-blue and violet feathers of *Cotinga cotinga* or *Cotinga amabilis*. The red and white feathers are taken from either *Pompadour cotinga* or *Trogon collaris*. The fan's green feathers come from Amazon parrots (*Amazona*, Psittacidae family). Yellow and red feathers are from the widely traded channel-billed toucan (*Ramphastos vitellinus*) as well as another, unidentified species. In addition to these American birds, the craftsperson also used feathers from the little egret (*Egretta garzetta*) to frame the fan. This common European wetland species featured prominently in the workshops of featherworkers in early modern Europe.[18] The depiction of two of these birds – the cotinga and/or trogon – on the central panel of the fan suggests that the artist had a wider interest in these birds' habitat and behaviours. This choice can also be understood as an iconographic pun on the artists' skill and artifice in the transformation of natural materials into an artefact, as the bird is shown cleansing the feathers just an early modern featherworker would have cleansed feathers prior to their processing.[19]

The identification of birds and feathers used to make featherwork can be a tricky venture, especially since comparable historical specimens often do not survive or, if they do, are in a desolate state resulting from storage conditions, insect damage and light exposure. Significant changes in the material properties of feathers also result from birds' changing habitats and nourishment over time. Under such trying circumstances, MSI can provide an additional entry point for informed

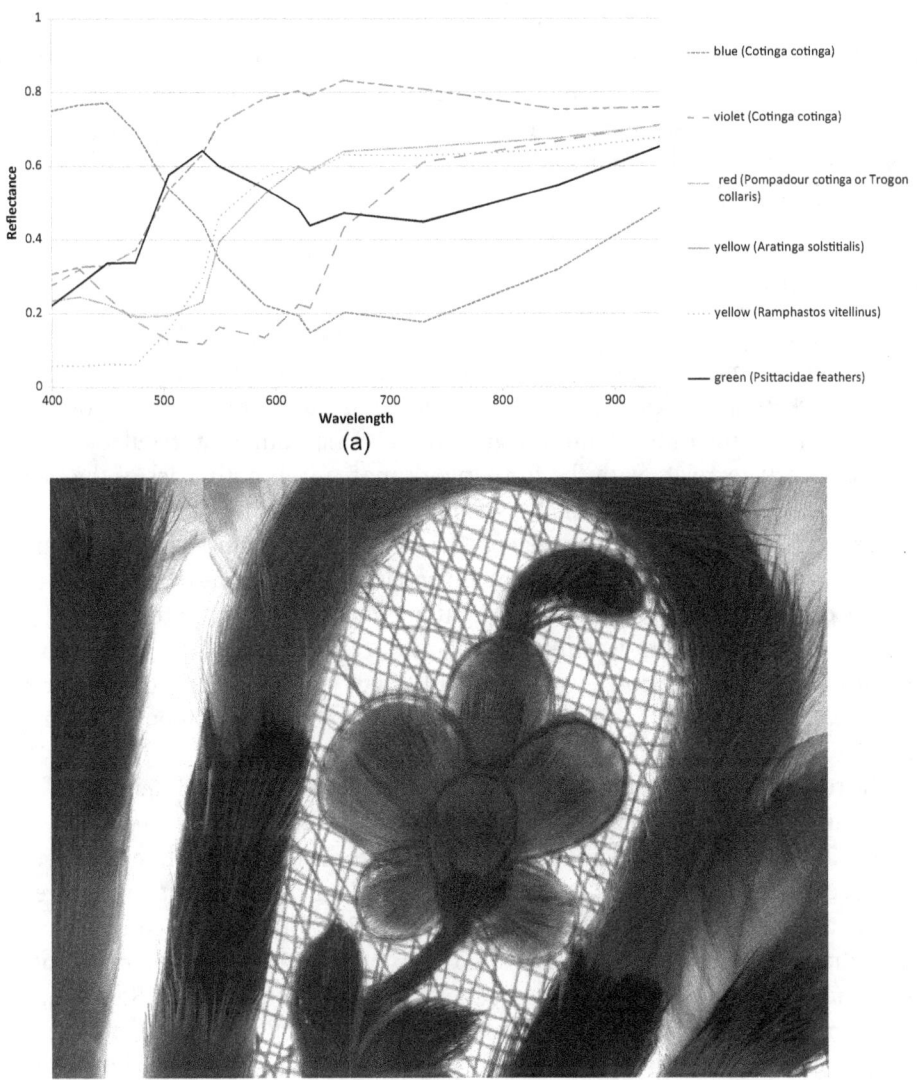

Figure 3 (a) Spectral image curves of identified feathers. © The John Rylands Research Institute and Library, University of Manchester. (b) Transmitted light imaging of the Messel Fan's threadwork panels, detailing the threads, felt templates with pinholes and the on-top layers of feathers. © The John Rylands Research Institute and Library, University of Manchester, reproduced by permission of the Fitzwilliam Museum, University of Cambridge.

contextualisation. As the iridescence of feathers is not visually stable, of course, spectral curve responses of feathers cannot be interpreted without contextualised research and comparison with specimens in particular. There remains an urgent need for more and better data, including wider datasets that contain historical samples, to better understand the impact of fading, ageing and conservation treatment. Research on the Messel Standing Feather Fan, however, shows the transformative potential of MSI analysis for more informed species identification in the future study of historical featherwork.

Crafts Knowledge

Advanced imaging also transforms our understanding of early modern feather crafts and the manual skills and material interventions of seventeenth-century featherworkers. The Messel Standing Feather Fan is made of five sections of drawn threadwork. Transmitted light imaging reveals that different regular schemes of threadwork correlate with the featherwork motives on both sides of the panels (Figure 4). Conducting such analysis from both sides of the fan shows the consciously arranged symmetry of visual patterns. Such work required intricate manual skills, complex calculations, prescient imagination, astute eyesight and strategic minute decision-making. Transmitted light imaging also provided further material insights. The threadwork was made of plant fibres that were twisted into double threads and not, as previously assumed, out of gut (Figure 3b). Scanning electron microscopic analysis would be required for further plant fibre identification, yet invasive sampling is not an option for this unique artefact.

Macrophotographs and focus stacking reveals that finely cut felt templates have been glued and stitched on to both sides of the threadwork panels. Feathers were then glued on to these felt templates in a layering technique that echoes the layered plumage of Amazonian birds and featherwork. Transmitted light imaging even reveals the pinholes covered by feathers (Figure 3b). Before they could be glued on to the felt, the artisan had to sort and cleanse the feathers from preen oil, using soap and hot water. The feathers were then cut into specific shapes. Some barbes, for example, have been cut into the shape of an insect's antennae (Figure 5a). Such interventions required the skilful handling of high-precision knives and tweezers at millimetre scale. Wrong handling could destroy the green iridescence of the barb that was essential to the fan's aesthetic appearance (Figure 5b). The feathers were cut according to their colours and then glued on to the templates to establish the fan's iconography of birds, flowers and butterflies (Figure 5b–c). Macrophotographs of insect damage reveal that at least six extremely thin layers of feathers were applied to create a butterfly motif (Figure 5a).[20]

In line with early modern understandings of artistic ingenuity, the featherworker clearly put effort into hiding traces of manual intervention. Feather layering concealed making interventions, and the targeted yet minimal application of adhesives is almost invisible to the naked eye. Advanced imaging, however, makes such interventions visible. Digital microscopy, macrophotography and MSI confirm the use

Figure 4 Transmitted light imaging of the Messel Fan. © The John Rylands Research Institute and Library, University of Manchester, reproduced by permission of the Fitzwilliam Museum, University of Cambridge.

of three different glues that show different colour/light responses and surface characteristics. The use of different adhesives corresponds with the different manual activities that had to be performed by the featherworker when working with materials. A golden, brownish adhesive was applied to fix the felt templates on to the threadwork (Figure 6a). A black, greyish glue with white, reflective increases was used to layer feathers (Figure 6b). This was a very special glue since classical adhesives would be expected to respond to near-infrared ranges above 1,200 nm.[21] This adhesive, however, is captured with a LP590 long pass red filter at 450nm (Figure 6d–e). A third viscous adhesive was used to knead feathers into distinct shapes (Figure 6c). This latter adhesive is reminiscent of resins from *Mimusops globosa* and *Moronobea coccinea*, which feature in Amazonian Indigenous featherwork, and which seventeenth-century Indigenous Amazonians traded with Europeans.

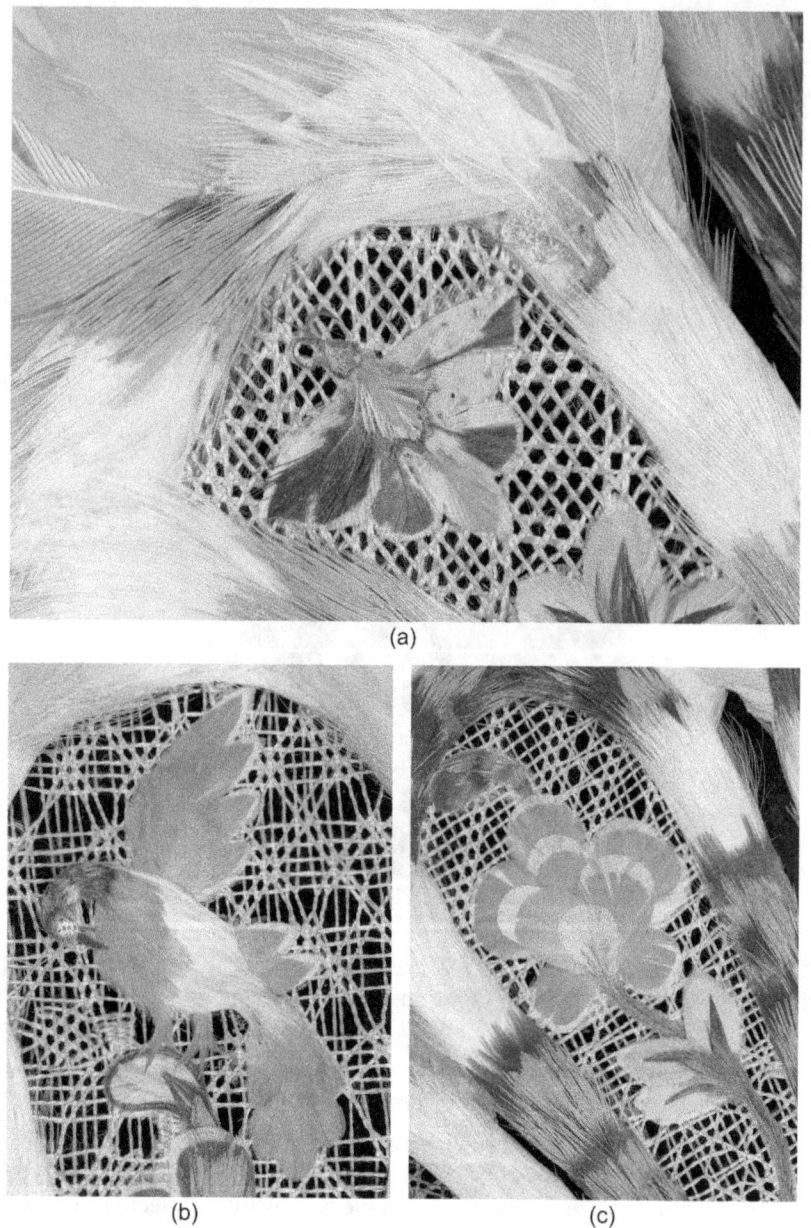

Figure 5 (a) Templates and layered featherwork exposed by insect damage. © The John Rylands Research Institute and Library, University of Manchester, reproduced by permission of the Fitzwilliam Museum, University of Cambridge. (b–c) Finely cut and layered feathers, glued into the iconographic shapes of flowers and birds. © The John Rylands Research Institute and Library, University of Manchester, reproduced by permission of the Fitzwilliam Museum, University of Cambridge.

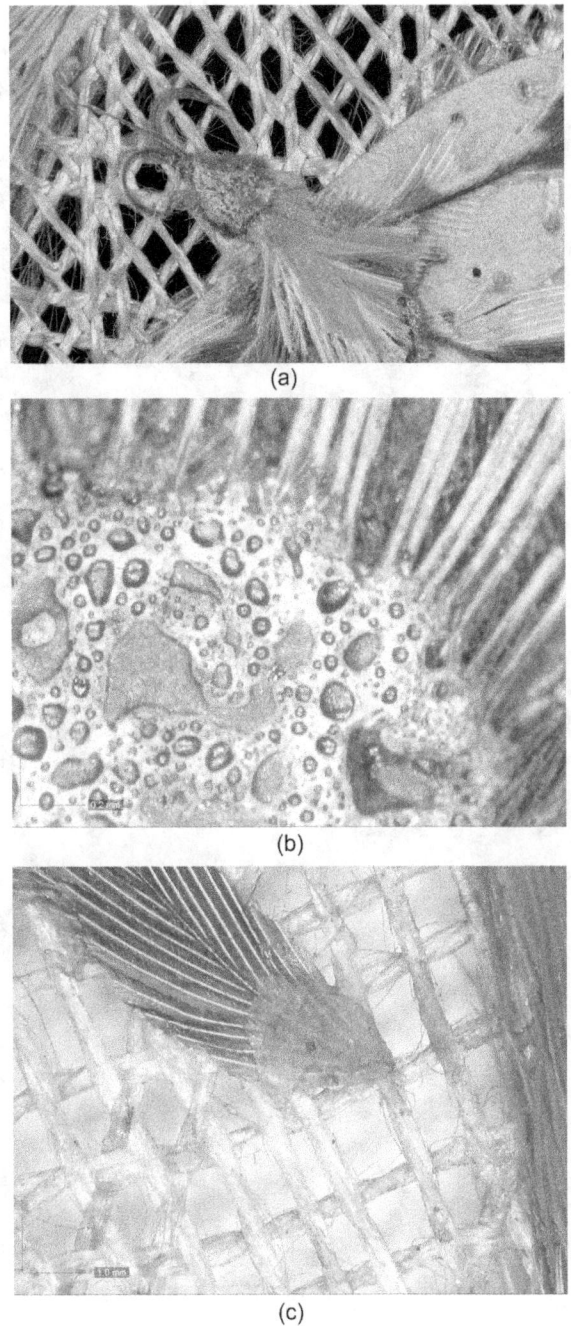

Figure 6 (a–b) Adhesive used for gluing feathers on feathers (detail of the cotinga's eye) and (c) adhesive used to knead layers of feathers into shapes, both seen under the Dino-Lite digital microscope. © Stefan Hanß.

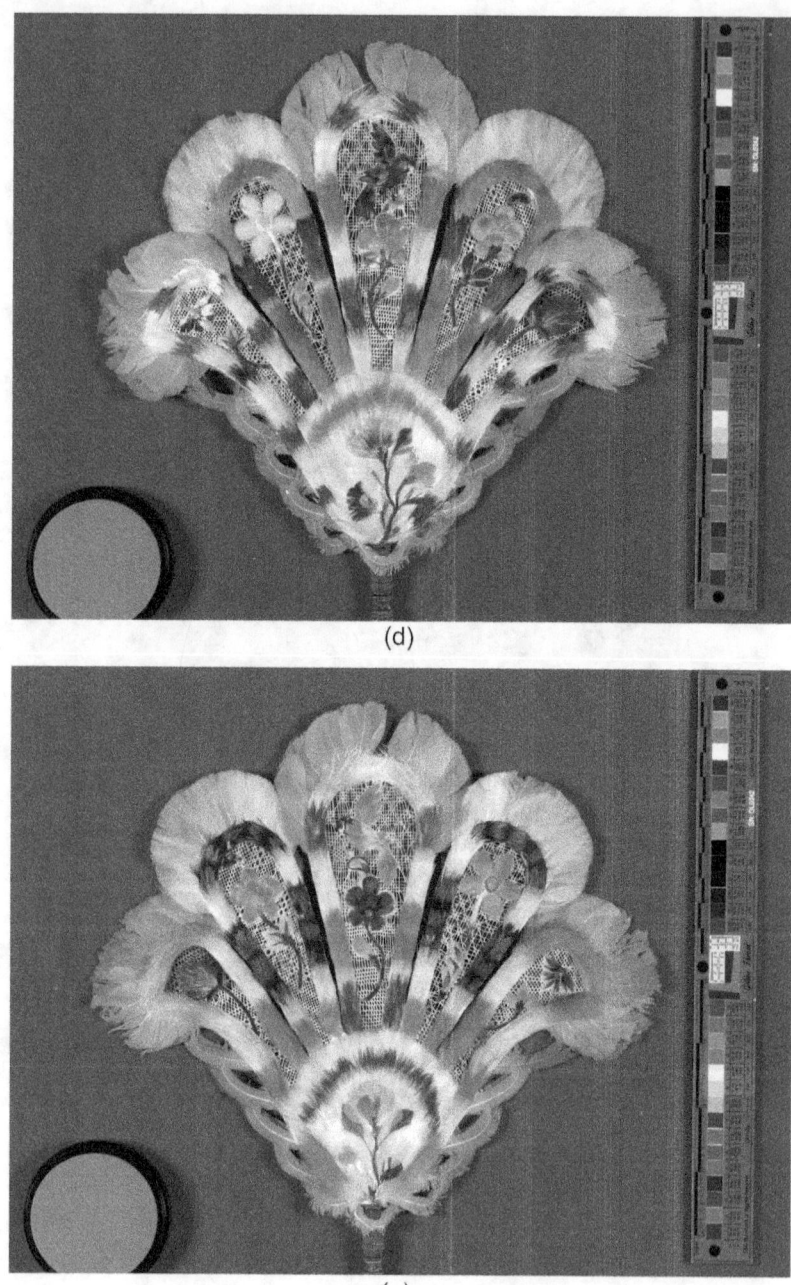

Figure 6 (d–e) Multispectral imaging (450nm, LP590 long pass red filter) with visual traces of fluorescent adhesive used to layer the feathers, and fluorescent traces of lost feathers. © The John Rylands Research Institute and Library, University of Manchester, reproduced by permission of the Fitzwilliam Museum, University of Cambridge.

Portuguese and French contemporaries wrote that these wax-like Amazonian resins (*yetic* and *yra-yetic*) resembled birdlime and kept away the moths.[22]

Advanced imaging thus unlocks some of the most minute and often invisible crafts interventions of featherworkers. We get a much fuller understanding of both the construction of the artefact and the choices of the artisan. Such imaging insights also lead to a greater appreciation of the complex material knowledge that went into the making of featherwork. Such artefacts materialise manual and cognitive achievements.

Conservation

Imaging is widely praised for rendering visible what is invisible to the human eye. In the case of featherwork, this comes with additional cultural significance. What we recover by studying past featherwork with advanced imaging techniques is in essence largely lost crafts knowledge. For Indigenous featherwork in particular we have the chance to recover culturally sensitive and often lost ancestral knowledge.[23] Future imaging research on Indigenous featherwork thus has the potential to be guided by descendant communities, responding to the needs of Indigenous practitioners and communities today to recover lost knowledge and grasp the ways in which Indigenous knowledge was transformed by processes of globalisation, environmental exploitation and colonisation. As research on this feather fan shows, Indigenous crafts cultures and Amazonian biodiversity transformed material aesthetics and practices around the world.[24]

Advanced imaging also partially recovers the appearance of the original artefact, thus providing a much better understanding of how the Messel Standing Feather Fan has changed over time. Today, the iridescence of its feathers is partially lost due to mechanical damage, preservation conditions, light exposure and ageing. High-resolution imaging, however, renders visible these feathers' stunning iridescent qualities. The fluorescent adhesive's response to LP590 long pass red filter at 450 nm, moreover, allows for the recovery of traces of lost featherwork through MSI (Figure 6d–e). The reflectance responses of the glue used to fix feathers on to feathers shows the silhouettes of lost feather layerwork. Moreover, macrophotographs reveal insect damage resulting from feathers' high level of keratin, which makes such materials prone to infestation. Historical insect damage can be studied with focus stacking, which has revealed unique insights into the layering techniques applied by the seventeenth-century artisan of the Messel Standing Feather Fan (Figures 5a, 7a). Focus stacking macrophotographs of insect damage even allowed for individual layers of feathers to be counted. We also identified insect larvae on the surface of feathers not visible with the human eye. Advanced imaging thus not only provides unique information about the history and making of this artefact but also essential information for its preservation for future generations.

Close microscopic analysis of the pale-pink-coloured feather frame of three panels of the Messel Standing Feather Fan shows colour variations that are characteristic of historical dyeing processes, pointing to the partial fading of dyestuff.

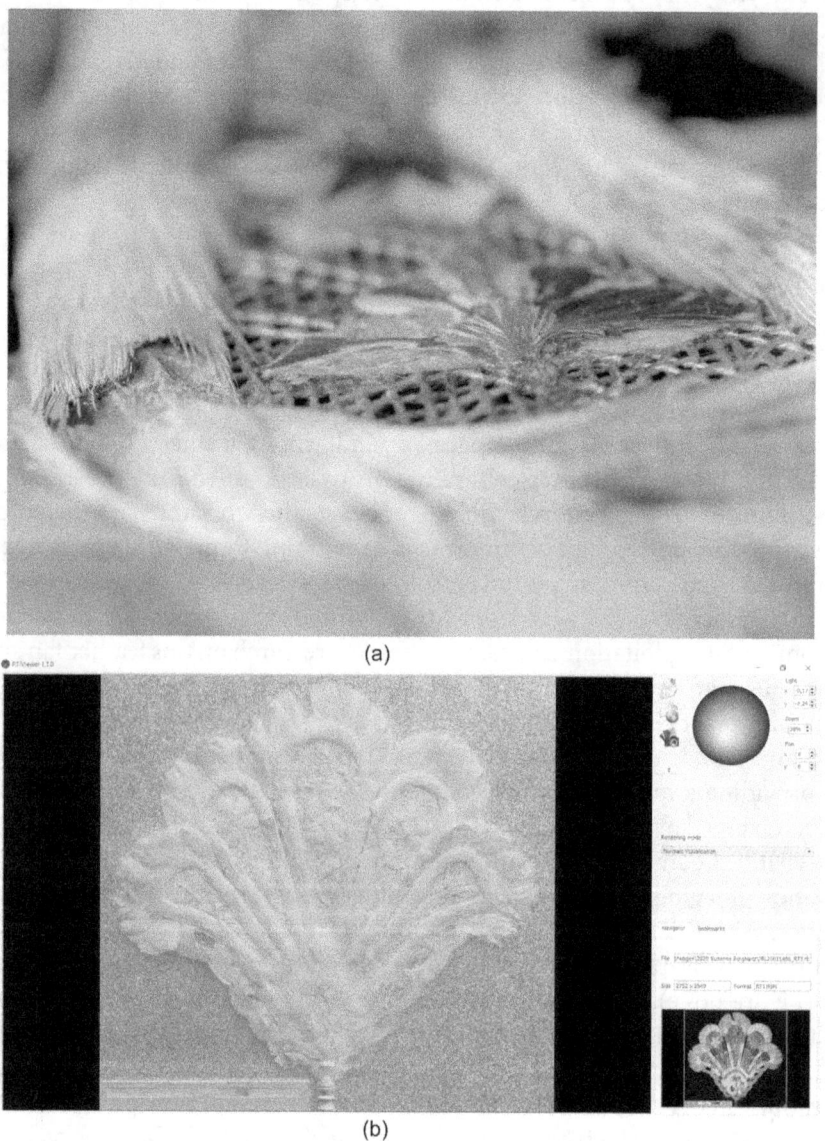

Figure 7 (a) Detailed view of two areas of insect damage. © The John Rylands Research Institute and Library, University of Manchester, reproduced by permission of the Fitzwilliam Museum, University of Cambridge. (b) Example screenshot of the RTI Viewer's visualisation of the Messel Standing Feather Fan. Normals Visualisation as seen here gives a rare and heightened impression of the fan's sensorial surface texture. The default rendering mode allows for the fan's iridescent response to different lighting options to be explored. See RTI file (https://drive.switch.ch/index.php/s/lFK9779qrwTGOSn [accessed 3 February 2025]) in open access RTI Viewer software (https://culturalheritageimaging.org/Technologies/RTI/ [accessed 3 February 2025]).

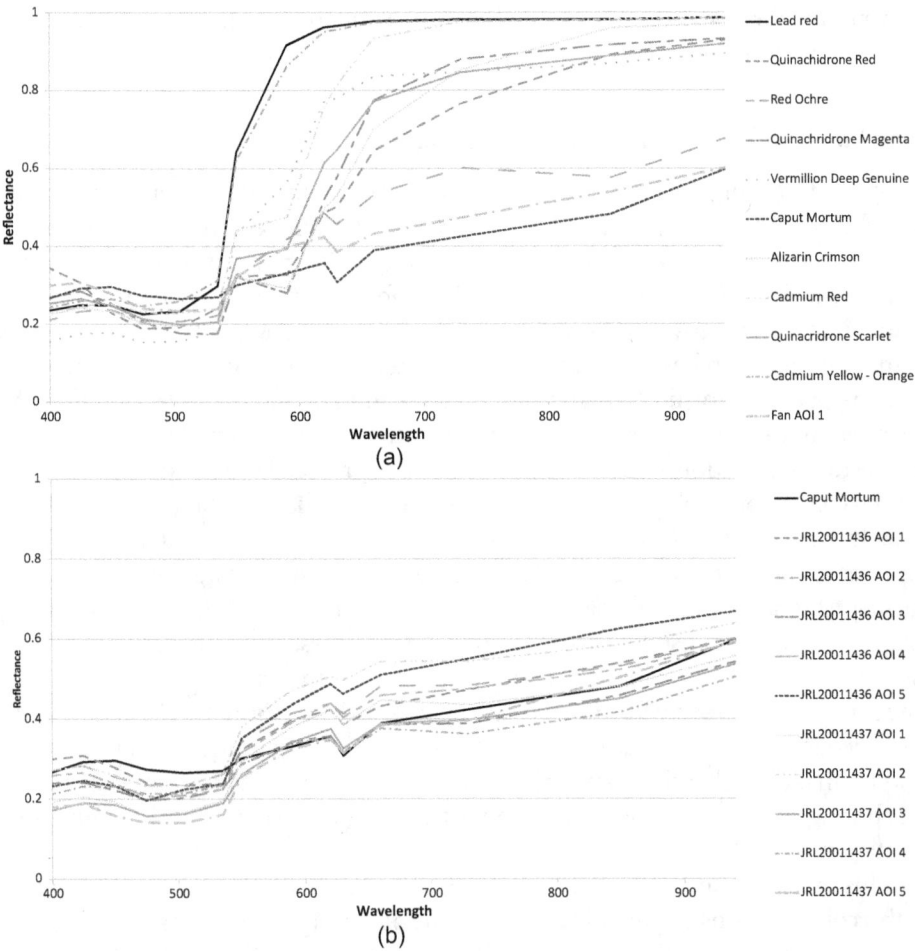

Figure 8 Spectral image curves of red pigments and ten areas of interest (AOIs) identified in the dyed feathers of both sides of the Messel Fan. Minor variations in 8b presumably result from varying reflectance of the surfaces of feathers. © The John Rylands Research Institute and Library, University of Manchester.

Early modern European artisans widely dyed feathers, archival documentation reveals, yet this knowledge was often secretive. While recipes for low-quality second-hand feather dyeing circulated widely, master featherworkers kept intricate dye recipes secret from the public and their competitors' prying eyes.[25] The dyestuff's spectral response matches the curve of *caput mortuum* – that is, burnt vitriol – an iron oxide pigment that was a common product of alchemical laboratory experiments at the time (Figure 8a–b).[26] Thus, the Messel Standing Feather Fan would not have been framed by white and pink feathers originally, but by white and brown feathers that echoed the wooden handle's colour and resonated with the contemporary European appreciation of brownish tints. Sixteenth-century Europeans, for instance, commented on the 'very special elegance' of American feathers of 'even [different] browns'.[27]

The Future of Featherwork Heritage Imaging

Advanced imaging approaches, as this discussion exemplifies, generate significant insights into the history of heritage featherwork. Yet this research also informs the future of feather artefacts, namely by increasing their preservation chances due to valuable insights for their conservation, like information on insect damage or dyestuff analysis. Moreover, RTI impacts the display of the Messel Standing Feather Fan. This approach has been applied to create a 3D surface model of the fan, which has the potential to substantially widen access to this unique artefact (Figure 7b). Featherwork ranks among the most fragile heritage; it is extremely vulnerable and prone to deterioration, dissolution, light sensitivity and insect damage. Considering the challenges of displaying such highly sensitive materials, RTI makes this artefact widely available to both visitors of exhibitions and users online. Filter and lighting options available in RTI Viewer software potentially transform audience experience as they generate a sense of what it means to handle such a fan.[28] This crucial sensory experience of object use comprises, among other factors, the feather fan's iridescent shimmers, reflectance capacity, colour vibrancy and the experience of listening to the sounds generated by the moving threadwork that could imitate the insects in the fan's iconography. Interactive museum and online display of this RTI-generated model makes such otherwise unavailable experiences accessible to audiences.[29] Such insights have the potential to impact exhibition designs and the online presentation of featherwork heritage in the future.

This article has situated the John Rylands Research Institute and Library Imaging Lab's collaborations within a wider landscape of heritage science research and heritage collection institutions to showcase the extent to which new interdisciplinary collaborations help develop innovative methodological approaches, and to highlight the expertise that library imaging scientists and photographers can bring to different materials. The methodological advancements discussed here are transferable and thus of major significance for featherwork heritage held in museum collections worldwide. Imaging featherwork methodology we argue, offers three major transformations for future research on featherwork: new knowledge, new data and new collaborations. First, this interdisciplinary approach shifts our understanding of featherwork across cultures, past and present, namely our knowledge of materials, techniques, crafts interventions and the knowledge of practitioners. Second, this methodological intervention comes with a call for better and more comprehensive data. It has been pointed out more generally 'that the documentation of current MSI applications to cultural heritage objects is often inconsistent, which means it is often difficult to interpret, compare, reuse, or reproduce results', and we therefore need FAIR – findable, accessible, interoperable and reusable – imaging data of heritage featherwork that allows for further insights and advancements.[30] Third, imaging heritage featherwork generates momentum for new collaborations to be built. Such collaborations could make past heritage relevant to shape a better future for Indigenous communities in facing today's challenges. Imaging methodologies, as discussed here, increase access for source communities and will hopefully open new, community-led conversations about the future display of featherwork heritage, the repatriation of featherwork and community-led collaborative heritage initiatives about the recovery of culturally sensitive featherwork knowledge responding to the needs of Indigenous communities and artist practitioners today.

Acknowledgements

The authors are extremely grateful to Vicky Avery (Fitzwilliam Museum, University of Cambridge), Jo Dillon (Fitzwilliam Museum, University of Cambridge), Paola Ricciardi (Natural History Museum London, formerly Fitzwilliam Museum, University of Cambridge), Gwen Riley Jones (Stockport Council, formerly John Rylands Research Institute and Library) and Helen Ritchie (Fitzwilliam Museum, University of Cambridge) for their support in bringing the collaboration, loan and analysis to fruition.

Notes

1 Christian Fischer and Ioanna Kakoulli, 'Multispectral and Hyperspectral Imaging Technologies in Conservation: Current Research and Potential Applications', *Reviews in Conservation*, 7 (2006), 3–16; Anna Pelagotti, Andrea Del Mastio, Alessia De Rosa and Alessandro Piva, 'Multispectral Imaging of Paintings', *IEEE Signal Processing Magazine*, 25:4 (2008), 27–36; Sotiria Kogou et al., 'A Holistic Multimodal Approach to the Non-Invasive Analysis of Watercolour Paintings', *Applied Physics A*, 121 (2015), 999–1014.

2 Joanne Dyer and Sophia Sotiropoulou, 'A Technical Step Forward in the Integration of Visible-Induced Luminescence Imaging Methods for the Study of Ancient Polychromy', *Heritage Science*, 5 (2017), article no. 24, https://doi.org/10.1186/s40494-017-0137-2.

3 Costas Papadopoulos, Yannis Hamilakis, Nina Kyparissi-Apostolika and Marta Díaz-Guardamino, 'Digital Sensoriality: The Neolithic Figurines from Koutroulou Magoula, Greece', *Cambridge Archaeological Journal*, 29:4 (2019), 625–52.

4 Joanne Dyer, Diego Tamburini, Elisabeth R. O'Connell and Anna Harrison, 'A Multispectral Imaging Approach Integrated into the Study of Late Antique Textiles from Egypt', *PLOS One*, 13:10 (2018): e0204699.

5 Rita Deiana, Alberta Silvestri, Manuela Gianandrea, Sarah Maltoni and Chiara Croci, 'The Medieval Glass Mosaic of S. Agnese fuori le mura in Rome: Multispectral Imaging for Preliminary Identification of Original Tesserae', *Heritage*, 6:3 (2023), 2851–62.

6 Wei-Ping Chan et al., 'A High-Throughout Multispectral Imaging System for Museum Specimens', *Communications Biology*, 5 (2022), article no. 1318.

7 Cerys Jones, Christina Duffy, Adam Gibson and Melissa Terras, 'Understanding Multispectral Imaging of Cultural Heritage: Determining Best Practice in MSI Analysis of Historical Artefacts', *Journal of Cultural Heritage*, 45 (2020), 339–50. For some of the 3D models produced by The Imaging Lab of the John Rylands Research Institute and Library, see https://sketchfab.com/TheJohnRylandsLibrary/models [accessed 13 February 2025].

8 Stefan Hanß, 'The Material Creativity of Affective Artifacts in the Dutch Colonial World: Imaging and Imagining Early Modern Feather Fans', *Current Anthropology*, 65:2 (2024), 196–234, including a discussion forum with Sven Dupré, Lydia Gibson, Yannis Hamilakis, Timothy J. LeCain, Lucio Menezes Ferreira, Ellen Pearlstein and Stefan Hanß.

9 Quotations from Cultural Heritage Imaging, 'Reflectance Transformation Imaging (RTI)', https://culturalheritageimaging.org/Technologies/RTI/ [accessed 7 October 2024], where the RTI Viewer can be downloaded.

10 Stefan Hanß, 'New World Feathers and the Matter of Early Modern Ingenuity: Digital Microscopes, Period Hands, and Period Eyes', in Richard J. Oosterhoff, José Ramón Marcaida and Alexander Marr (eds), *Ingenuity in the Making: Materials and Technique in Early Modern Europe* (Pittsburgh: University of Pittsburgh Press, 2021), pp. 189–202, 195; Stefan Hanß, 'Material Encounters: Knotting Cultures in Early Modern Peru and Spain', *The Historical Journal*, 62:3 (2019), 583–615; Stefan Hanß, 'Digital Microscopy and Early Modern Embroidery', in Anne Gerritsen and Giorgio Riello (eds), *Writing Material Culture History* (London: Bloomsbury, 2nd edn, 2021), pp. 214–21.

11 Julien Vermeulen, Stefan Hanß and Ulinka Rublack, '"In Feather-Working, the Only Limitation is One's Own Imagination": Julien Vermeulen, Stefan Hanß and Ulinka Rublack on the Past and Future of Feather-Working', *Microscopic Records: The New Interdisciplinarity of Early Modern Studies, c.1400–1800*: 21 October 2019, https://sites.manchester.ac.uk/microscopic-records/2019/10/21/in-feather-working-the-only-limitation-is-ones-own-imagination-julien-vermeulen-stefan-hans-and-ulinka-rublack-on-the-past-and-future-of-feather-working/ [accessed 13 February 2025].

12 Nathan S. Hart, 'Variations in Cone Photoreceptor Abundance and the Visual Ecology of Birds', *Journal of Comparative Physiology A*, 187 (2001), 685–97; Nathan S. Hart and David M. Hunt, 'Avian Visual Pigments: Characteristics, Spectral Tuning, and Evolution', *American Naturalist*, 169:1 (2007), S7–S26.

13 Kathryn E. Arnold, Ian P. F. Owens and N. Justin Marshall, 'Fluorescent Signaling in Parrots', *Science* 295:5552 (2002), 92.

14 Jan Tinbergen, Bodo D. Wilts and Doekele G. Stavenga, 'Spectral Tuning of Amazon Parrot Feather Coloration by Psittacofulvin Pigments and Spongy Structure', *Journal of Experimental Biology*, 216 (2013), 4358–64; Ilke Van Hazel, Amir Sabouhanian, Lainy Day, John A Endler and Belinda S. W. Chang, 'Functional Characterization of Spectral Tuning Mechanisms in the Great Bowerbird Short-Wavelength Sensitive Visual Pigment (SWS1), and the Origins of UV/Violet Vision in Passerines and Parrots', *BMC Evolutionary Biology*, 13 (2013), 1–13; José M. Medina, José A. Díaz and Pete Vukusic, 'Classification of Peacock Feather Reflectance Using Principal Component

Analysis Similarity Factors from Multispectral Imaging Data', *Optics Express*, 28:8 (2015), 10198–212; Suzanne Amador Kane, Yuchao Wang, Rui Fang, Yabin Lu and Roslyn Dakin, 'How Conspicuous are Peacock Eyespots and Other Colorful Feathers in the Eyes of Mammalian Predators?', *PLOS One*, 14:4 (2019), 1–35.

15 Renée Riedler, Christel Pesme, James Druzik, Molly Gleeson and Ellen Pearlstein, 'A Review of Color-Producing Mechanisms in Feathers and their Influence on Preventive Conservation Strategies', *Journal of the American Institute for Conservation*, 53:1 (2014), 44–65.

16 Ellen Pearlstein et al., 'Ultraviolet-Induced Visible Fluorescence and Chemical Analysis as Tools for Examining Featherwork', *Journal of the American Institute for Conservation* 54:3 (2015), 149–67; Ellen Pearlstein (ed.), *Conservation of Featherwork from Central and South America* (London: Archetype, 2017).

17 Sylvie Colinart, T. Borel, C. Moulherat, E. Ravaud, C. Chahine and F. Juchaud, 'Le triptych de la crucifixion sous l'œil du laboratoire', in Alain Erlande-Brandenburg (ed.), *Le triptyque aztèque de la crucifixion, Paris: réunion des musées nationaux* (Écouen and Paris: Musée national de la Renaissance, 2004), pp. 102–8; Alessandra Russo, *The Untranslatable Image: A Mestizo History of the Arts in New Spain* (Austin: University of Texas Press, 2014); Alessandra Russo, Gerhard Wolf and Diane Fane (eds), *Images Take Flight: Feather Art in Mexico and Europe, 1400–1700* (Munich: Hirmer, 2015); Ellen J. Pearlstein, 'Bishop's Miter and Infulae, a Feathered Masterpiece from Museo degli Argenti in Florence', *Latin American and Latinx Visual Culture*, 1:2 (2019), 99–106.

18 See the workshop inventories published in Stefan Hanß, 'Making Featherwork in Early Modern Europe', in Susanna Burghartz, Lucas Burkart, Christine Göttler and Ulinka Rublack (eds), *Materialized Identities in Early Modern Culture, 1450–1750: Objects, Affects, Effects* (Amsterdam: Amsterdam University Press, 2021), pp. 137–85.

19 *Ibid.*, pp. 163, 169–77.

20 On the respective steps of featherworking, see Hanß, 'Making Featherwork'; Hanß, 'Material Creativity'; Vermeulen, Hanß and Rublack, '"In Feather-Working, the Only Limitation is One's Own Imagination"'.

21 Sotiria Kogou et al., 'The Origins of the Selden Map of China: Scientific Analysis of the Painting Materials and Techniques using a Holistic Approach', *Heritage Science*, 4 (2016), 1–24, 6.

22 Jean De Léry, *History of a Voyage to the Land of Brazil, Otherwise called America*, trans. Janet Whatley (Berkeley: University of California Press, 1990), pp. 60, 92; Donald W. Forsyth, 'The Beginnings of Brazilian Anthropology: Jesuits and Tupinamba Cannibalism', *Journal of Anthropological Research*, 39:2 (1983), 147–78, 164–65, 178; Claude Lévi-Strauss, 'The Use of Wild Plants in Tropical South America', in Julian H. Steward (ed.), *Handbook of South American Indians*, vol. 6 (New York: Cooper Square, 1963), pp. 465–86, 477; Warren Dean, *With Broadax and Firebrand: The Destruction of the Brazilian Atlantic Forst* (Cambridge, MA: Harvard University Press, 1997), p. 130.

23 Paul Sillitoe, 'The Development of Indigenous Knowledge: A New Applied Anthropology', *Current Anthropology*, 39 (1998), 223–52; Aileen Moreton-Robinson,

Critical Indigenous Studies: Engagements in First World Locations (Tucson: University of Arizona Press, 2016).

24 Hanß, 'Material Creativity'.
25 On feather dyeing, Ulinka Rublack, 'Befeathering the European: The Matter of Feathers in the Material Renaissance', *The American Historical Review*, 126:1 (2021), 19–53; Hanß, 'Making Featherwork', pp. 169–70; Stefan Hanß, 'Feathers and the Making of Luxury Experiences at the Sixteenth-Century Spanish Court', *Renaissance Studies*, 37:3 (2023), 399–438.
26 Nicholas Eastaugh, *Pigment Compendium: A Dictionary and Optical Microscopy of Historical Pigments* (London: Routledge, 2013), p. 87; Sven Dupré (ed.), *Laboratories of Art: Alchemy and Art Technology from Antiquity to the 18th Century* (Cham: Springer, 2014); Spike Bucklow, 'Alchemy and Colour', in Stella Panayotova (ed.), *Colour: The Art and Science of Illuminated Manuscripts* (London: Harvey Miller, 2016), pp. 108–17, 112–13. We used JRRIL 16070880 red pigment sample collection as reference.
27 Peter Martyr, *De orbe novo decades I–VIII*, eds Rosanna Mazzacane and Elisa Magoncalda (Genova: Dipartimento di Archaeologia, Filologia classica e loro tradizioni, 2005), 4:9, 16–17.
28 The RTI file can be downloaded at https://drive.switch.ch/index.php/s/lFK9779qrwTGOSn [accessed 3 February 2025] and explored via the open access RTI Viewer software available at https://culturalheritageimaging.org/Technologies/RTI/ [accessed 3 February 2025].
29 Papadopoulos, Hamilakis, Kyparissi-Apostolika and Díaz-Guardamino, 'Digital Sensoriality'.
30 Jones, Duffy, Gibson and Terras, 'Understanding Multispectral Imaging', 340.

Sri Guru Granth Sahib: Collaboration, Digitisation, Heritage and the Legacy of Colonial Collections

GURTEK SINGH, UNIVERSITY OF MANCHESTER
JAMES ROBINSON, UNIVERSITY OF MANCHESTER

Abstract

This article examines the contested legacy of colonial collections, focusing on the digitisation and dialogical curation of the Sri Guru Granth Sahib held at the John Rylands Library. It explores the ethical, legal and cultural implications of preserving and digitising sacred artefacts acquired through colonial exploitation. Using the example of collaboration between the University of Manchester and the Sikh community, the article highlights how digital technologies can both democratise access and perpetuate epistemological violence. By tracing the history, provenance and recent community consultations, the article underscores the importance of dialogical curation in addressing the legacies of colonialism.

Keywords: Sikh heritage; Sri Guru Granth Sahib; colonialism; digitisation; dialogical curation; community engagement

Introduction

The presence of manuscripts such as the Sri Guru Granth Sahib, or Punjabi MS 5,[1] in institutions like the John Rylands Library prompts urgent questions about the enduring legacies of colonialism and the ethical responsibilities tied to cultural heritage preservation, digitisation and efforts towards decolonisation and repatriation. These artefacts, often acquired through colonial violence or exploitation, carry with them histories that demand critical interrogation of how they are housed, displayed and represented today.

The poem *Here, and Here, and Here*, by Suhaiymah Manzoor-Khan, serves as a powerful critique of the colonial legacies embedded within the John Rylands Library itself. Commissioned by the library for the exhibition *We Have Always Been Here*, the poem confronts the uncomfortable truths of the library's origins. Manzoor-Khan's research into the collections and the library's historical foundations – built on wealth derived from the cotton industry and the transatlantic slave trade – inspired the piece. In her own words, 'There is no Rylands without slavery and plantation; and there is no Crawford collection without colonisation.'[2]

In an article written for the John Rylands Library blog, Manzoor-Khan reflects further on the physical and historical weight of the building and its collections, and how this shaped her response:

> I could leaf through the letters written by colonial colonel, Samuel Bagshaw, when he was part of the East India Company; but to find the voices of the people being colonised and brutalised was much harder. I could browse digitised collections of Quran's, Persian scripts, Rumi's original Masnavi, and more; but the secular lens through which such manuscripts were narrated or curated meant I felt I was missing the reality of those scripts, the hearts of the worshippers who calligraphed and wrote them.[3]

These powerful reflections situate the Rylands not as a neutral repository of knowledge but as a site entangled with histories of exploitation, colonial conquest and systemic erasure. They raise essential questions about how institutions can move beyond merely acknowledging these legacies to actively addressing the epistemological violence (the act of silencing marginalised groups by denying the validity of their knowledge systems)[4] inherent in their foundations and practices. Through works like Manzoor-Khan's poem, we are reminded that to decolonise these institutions is not just to revisit their histories but to rethink their role in the present and their obligations to the communities whose cultures and histories they hold.

The *Sikh Heritage Project* was partly born out of a public lecture in August 2021 titled 'Exploring Manchester's Histories: Sikhs in the City'.[5] The session was hosted by Gurtek Singh and Maya Sharma, collections engagement officer at the Ahmed Iqbal Ullah RACE Centre[6] (part of the University of Manchester) and consultancy manager for the Ahmed Iqbal Ullah Education Trust.[7] The centre is focused on collecting with Global Majority communities and has a long-standing practice of sustained community engagement. The session explored the roots of early Sikh migration to the city and the influence the communities had on the region. The items used for this session came from the Sikh Family History Project collected in 1982. While the session was a success, it highlighted areas which were deficient in research.

In the 1980s and 1990s there was long a rumour within the Manchester Sikh community that there existed a version of the Sri Guru Granth Sahib held within the University of Manchester. This was, of course, true, and in 2004 a request was made by local Sikhs to access it at the library. While the request was granted, such was the suspicion towards the community a security team was put on stand-by for the visit. Dr John Hodgson began working at the Rylands in 1989 as an archivist and retired in 2024 from the role of associate director. When reflecting on how attitudes had changed, Hodgson commented: 'Twenty-five years ago, our approach was characterised by ignorance of the Sri Guru Granth Sahib's religious and cultural significance for Sikhs; embarrassment over its implication in British colonialism and militarism; and suspicion of the motives of any members of the community who asked to see it.'[8]

In 2021, the university's approach to engaging with the Sikh community began to evolve. The success of the 'Sikhs in the City' session demonstrated strong community interest and sparked enquiries about other Sikh heritage items within the university. In response, Maya Sharma arranged an initial consultation with Gurtek Singh and Roopa Kaur Singh, a community member fluent in Gurmukhi, leading to a proposal to host community groups on campus. The RACE Centre then tasked Dr Inbal Livne, the incoming curator of diversifying collections at the Rylands, with facilitating access arrangements.

Trust between the university and the community developed gradually over several years, strengthening relationships and fostering meaningful collaboration. This engagement brought renewed attention to public outreach and decolonisation efforts, paving the way for more inclusive initiatives.

Provenance and History

Undoubtedly, the most significant item to the Sikh community is the Sri Guru Granth Sahib. The 'Guru'[9] is the spiritual guide in the Sikh faith. Sri Guru Nanak Dev Ji (1469–1539) was the first of ten Gurus and resided within the Punjab region of South Asia. Manuscripts of the Gurus, saints and poets were collected and compiled together and in 1604 installed in Amritsar by Guru Arjan Dev Ji.[10] Max Arthur Macauliffe (1838–1913) wrote the first English translation of the Sri Guru Granth Sahib, which was endorsed by Sri Akal Takht Sahib, the highest Sikh body. Macauliffe states 'the line of the Gurus ended with the tenth, Guru Gobind Singh: 'He ordered that the Granth should be to his Sikhs as the living Gurus'.[11] Basics of Sikhi, a leading and influential Sikh educational charity in the UK, explain: 'Guru Granth Sahib is the universal eternal Guru of the Sikhs. Guru Granth Sahib Ji is the holy scripture of Sikhs that is treated with utmost respect and is respected by Sikhs as a living Guru and not merely a scripture.'[12] The Granth is considered to be a living Guru and so is treated in a manner of a living person of the highest social status.

As with many other collection items, the history of the Guru in the Rylands is documented only from the moment it became a colonial acquisition. Its journey to the library is entwined with the broader narrative of empire and the voracious appetite of the colonial elite for rare and 'exotic' texts, often obtained through dubious or exploitative means. The Bibliotheca Lindesiana, purchased by Enriqueta Rylands in 1901, now housed in the John Rylands Library, was built over the latter half of the nineteenth century by Alexander Lindsay and his son James Ludovic Lindsay, the 25th and 26th Earls of Crawford. They amassed over 3,000 'oriental' books and manuscripts,[13] including the Guru, reflecting the era's fascination with exoticism, imperial conquest and the obsessive collecting trends of 'bibliomania' that marked the late eighteenth and nineteenth centuries.[14]

The acquisition of the Guru exemplifies the violent and coercive underpinnings of such collections. In June 1862, Bernard Quaritch, a London bookseller

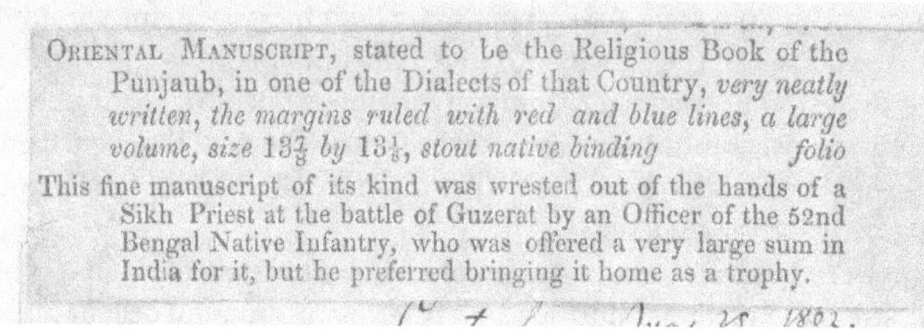

> ORIENTAL MANUSCRIPT, stated to be the Religious Book of the Punjaub, in one of the Dialects of that Country, *very neatly written, the margins ruled with red and blue lines, a large volume, size* 13¾ *by* 13½, *stout native binding*   *folio*
> This fine manuscript of its kind was wrested out of the hands of a Sikh Priest at the battle of Guzerat by an Officer of the 52nd Bengal Native Infantry, who was offered a very large sum in India for it, but he preferred bringing it home as a trophy.

Figure 1 Entry from the sale in the 'Handlist of Hindustani, Marathi & Punjabi Manuscripts'. © The John Rylands Research Institute and Library, University of Manchester.

specialising in such manuscripts, wrote to Lindsay to offer what he described as an exceptional and singular acquisition: 'I have just secured a most extraordinary manuscript, a perfect copy of the Sacred book of that gallant nation the Sikhs called the "Grunth." During all my experience as an oriental bookseller, I never had such a manuscript before.'[15]

The Guru was noted in the catalogue of sale as having been 'wrested out of the hands of a Sikh Priest at the Battle of Guzerat (Gujrat) on February 21, 1849, by an Officer of the 52nd Bengal Native Infantry' (Figure 1).[16] This stark language highlights the violent origins of its acquisition during the Second Anglo-Sikh War, which marked the final annexation of Punjab into the British Empire. The Guru was thus treated as a 'trophy', reflecting the colonial tendency to decontextualise sacred items and reframe them as commodities.

Lindsay, demonstrating no regard for the circumstances of its obtainment, deemed the manuscript to be of 'great curiosity & value' and willingly paid £21 for it.[17]

In 1901, the Bibliotheca Lindesiana was sold to Enriqueta Rylands for the substantial sum of £155,000. For decades however, the manuscript appeared to be largely ignored, and evidence suggests that requests to access the Guru were often met with 'fear and suspicion',[18] revealing a deep disconnect between the library's custodianship and the manuscript's sacred status.

The origins and implications of such acquisitions were deliberately obscured, aligning with efforts to suppress evidence of their connection to colonial violence and exploitation. Recent efforts to uncover and address these histories have shed light on the complex ethical and historical issues surrounding such items, raising important questions about restitution, cultural sensitivity and the responsibilities of modern custodianship, and the threat of continued epistemological violent practice.

First coined by theorist Gayatri Chakravorty Spivak in 1988, 'epistemological violence' is a critical aspect of knowledge institutions' ongoing role in confronting the legacy of colonialism.[19] This includes addressing not only the colonial acquisition and misrepresentation of objects but also the epistemic injustices embedded in the structures and practices of knowledge institutions. In 'Museums and the Epistemology of Injustice: From Colonialism to Decoloniality' Shadid Vawda emphasises the need to acknowledge and dismantle these injustices, which are perpetuated when institutions fail to critically engage with the coloniality of their collections, exhibitions and interpretive frameworks.[20] Yirga Gelaw Woldeyes, in his work on Ethiopian manuscripts, expands on this by illustrating how African heritage is often interpreted through a Eurocentric lens, resulting in the internalisation of colonial frameworks within local communities themselves. This process reinforces epistemological hierarchies, silencing Indigenous perspectives and reframing local knowledge.[21]

The first significant academic reference to the Guru in the Rylands appears in Pashaura Singh's *The Guru Granth Sahib: Canon, Meaning and Authority* (2003). Singh recognised it as the earliest known manuscript of the Guru Granth Sahib held outside India.[22] That same year, a Rumalla Sahib (a ceremonial covering) was donated to the library, marking an emerging awareness of the manuscript's sacred nature. However, despite this recognition, staff at the time reportedly hesitated to share details about its provenance or storage conditions.

The Sikh Heritage Project

In the years leading up to the current project, consultations with Sikh scholars gradually increased awareness of the manuscript's importance. Various funding bids were submitted to support its conservation, reflecting growing recognition of the need for both preservation and ethical stewardship. However, none materialised.

In 2008, conservator Mark Furness produced a detailed condition report that explicitly acknowledged the Guru Granth Sahib as a living Guru. The report's introduction raised critical questions about the manuscript's ownership and whether 'there is a moral imperative to return the manuscript'. While the proposed treatment plan was relatively standard, it notably recommended looking for examples of best practices in conserving Sikh manuscripts, signalling a shift towards more culturally sensitive approaches (Figure 2).[23]

It was not until 2020 that substantial work on the Guru began, with pre-conservation discussions and planning between the community and lead conservator Laura Snow, marking a decisive step towards addressing physical preservation and the broader ethical questions tied to its colonial history. Conservation work was carried out over a six-month period at the Rylands, in a dedicated space adhering to Sikh practices.

The Sikh community was first established in Manchester in the 1930s. The 2021 census accounted that those who identified as a 'Sikh' in Greater Manchester numbered 7,350.[24] This figure was nearly double the 2001 population of 3,720.[25]

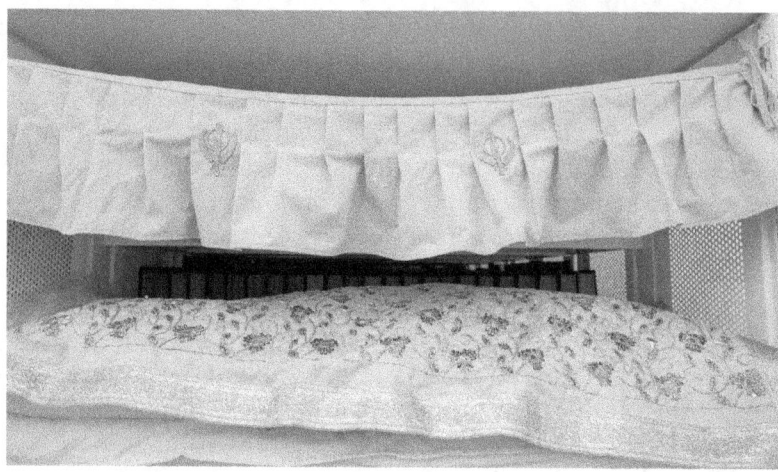

Figure 2 The Sri Guru Granth Sahib, now stored in line with Sikh tradition, wrapped in one of a number of community-donated Rumalla Sahibs. © The John Rylands Research Institute and Library, University of Manchester.

Despite this increase, the number of Gurdwaras[26] remained at five. This meant that their leadership was long established, and they represented significant numbers of the local Sikh population. The five Gurdwaras collectively organised events across Manchester including the annual parade across the city known as the Manchester Nagar Kirtan. The project greatly benefited from the well-established infrastructure within Manchester's Sikh community.

Consultation for the project involved several meetings between the university and Gurdwara representatives, held both at the Rylands and Gurdwaras from August 2022. Representatives paid respects to the Sri Guru Granth Sahib as well as Sri Dasam Granth (Punjabi MS 6)[27] and a historic Pothi (Punjabi MS 4).[28] The sessions at the library were held in a dedicated room on the highest floor which was prepared appropriately. Passages from the Guru were sung, and a Rumalla Sahib, a cloth covering, was symbolically placed on the Guru (Figure 3). This is an example of the Rylands adapting its consultation policy to be more appropriate and respectful. Further consultation sessions included the welcoming of a senior Granthi, a guardian of the Guru from Sri Darbar Sahib[29] in August 2023, and representatives of the Nihung Akaali Sikhs in June 2024. The consultation process included perspectives of the local community, the national Sikh population and international Sikh organisations.

Community opinion was complex. While the community overwhelmingly supported the digitisation of Sikh heritage in the region, perspectives differed on the Sri Guru Granth Sahib. Some in the community wholly disagreed with the storing of the Guru in an archive. Their position centred on the

Figure 3 A Senior Granthi from Sri Darbar Sahib (centre) and delegates from Sri Anandpur Sahib consult the historic Guru Granth Sahib seated on a Palki Sahib, donated to the Rylands by the community. Image: Gurtek Singh.

controversial provenance and the moral obligation of repatriation of the Guru to a natural setting in a Gurdwara. Others, while not comfortable with idea of the Guru in an archive, recognised the role the university had played in preserving such a significant item. It was deemed important that the university was progressive and actively engaged with community groups. Many believed the university had demonstrated positive intentions and could be a beacon to other institutions and collectors who withhold access, and therefore they supported the project.

A challenge of the project was the nuanced relationship and balance of power between the university and Gurdwaras. Many within the community perceived the university as the ultimate gatekeeper: heritage was held at the university, and it had the authority to control present and future access. The university on the other hand was aware of the influence of the Gurdwaras and other Sikh organisations on public opinion. For everyone involved in the project, it was therefore important that the consultation was a two-way relationship. The university was offering its skills and expertise while, equally, the community was offering its subject knowledge and networking. The community recognised the importance of digitisation but acknowledged the challenges with costs and expertise. Community members also agreed that conservation was a prerequisite for digitisation, and physical access to the Guru required a stable condition. Collaboration among all parties was crucial to define their goals and identify the most suitable approach to move the project forward.

Dialogical Curation

Dialogical curation is a way to challenge the colonial framing of institutional practices. This approach involves active engagement with Indigenous and diasporic communities, prioritising their voices, values and perspectives. Dialogical curation rejects the traditional, singular narrative often imposed by knowledge institutions and instead fosters dynamic exchanges between curators, scholars, communities and the public. Through this method, the process of curation becomes a collaborative endeavour that acknowledges the multiplicity of meanings and significances attached to cultural artefacts.[30]

In post-colonial contexts such as in Canada and South Africa, dialogical curation and decolonial practices have led to significant changes in how knowledge institutions address issues of ownership, representation and repatriation. Leanne Unruh highlights the importance of Indigenous self-representation as a cornerstone of these efforts, enabling communities to reclaim their narratives and challenge institutional power dynamics.[31] For contested cultural objects in Western institutions, such as the Guru, or Ethiopic manuscripts, adopting similar models could facilitate meaningful institutional change.

The community consultation process was deemed to have reached a point of sufficient engagement and planning to support the creation of a robust methodological framework. A key component of the broader project involved hosting a digital exhibition. This initiative aimed to:

- foster community engagement by providing a platform for individuals to share their thoughts and ideas about specific items and the digital archive;
- establish an accessible online resource where the Sikh community could explore heritage materials held at the university;
- encourage the community to contribute its invaluable subject knowledge to enhance understanding and stimulate broader discussions surrounding these historic artefacts.

The digital exhibition was organised around three primary themes, the first focusing on the digitisation of Sikh religious heritage, including significant texts such as the Sri Guru Granth Sahib and the Sri Dasam Granth.

The second addressed the digitisation of Sikh and Punjabi cultural artefacts encompassing arts, literature and weaponry. Key items included Punjabi MS 1, a manuscript of Punjabi poems,[32] and Punjabi MS 3, a traditional romantic novel (*Heer Ranjha*),[33] both held in the Rylands. The exhibition also includes artefacts from Manchester Art Gallery, highlighting the breadth of Sikh heritage preserved across the city.

The third strand of the exhibition focused on responding to community concerns about the risk of contemporary Sikh heritage being lost. The initiative sought to document and preserve stories already from within the community but not formally recorded. This work led to a collaborative effort between the Rylands, the

RACE Centre and the Education Trust to support new collecting practices. A digitisation day was held at Sri Guru Gobind Singh Gurdwara in Whalley Range in June 2023, where Sikh community members were invited to bring personal or familial items of cultural significance for careful documentation. Crucially, the primary goal was to build a collection housed at the RACE Centre, with the online platform serving as a secondary outcome. Digital donation discussions were central to the process, ensuring that community members retained full control over how their contributions were used, with no obligation for items to be made publicly accessible.

The collaboration was a new way of working for all involved, combining the Rylands' digital expertise with the Education Trust's established approach to community-led engagement. This exchange strengthened relationships and underscored the importance of centring community interests. The success of the event relied on trust-building, and much of this was made possible through the guidance of Gurtek Singh, who played a crucial role in liaising with the Gurdwara committees and wider community members. His careful facilitation ensured the process was respectful, culturally informed and rooted in community priorities, demonstrating the value of working in partnership to document and protect Sikh heritage in Manchester.

Benefits

A significant benefit of the conservation and digitisation of Punjabi MS 5 was that it allowed for a much greater understanding of the Granth. It was evident that the text had been modified on multiple occasions through the visible amendments, with some passages being concealed with a yellow paste called *hartal*, and corrections added in margins. An analysis of the contents suggests it may have been created in the late seventeenth or early eighteenth century during the period of the Tenth Guru (1666–1708).[34] The Granth could also be compared with other recensions that could improve the wider understanding of historic Guru Granth Sahibs and how they have evolved over time. The digitisation aids researchers to delineate where Sikh heritage resides across not just the UK but globally. They could also study the accompanying provenance documentation, draw links and fill knowledge gaps between items with a shared heritage.

The presence of passages mostly towards the end of this recension of the Guru caused challenges. As early as 2003, and throughout the consultation sessions, it was apparent that there are subtle differences between this Granth and the standardised printed version found in Gurdwaras across the UK. Written long before the standardisation, there are passages which do not feature in the contemporary version, including the *Chalitar Jothi Joth Ka*, which details when each of the Ten Gurus passed from the world, *Rattanmala Mahala Pehla*[35] and *Haqiqat Rah Muqam Raje Shivnabh Ki*.[36] After *Raagmala*, the Granth ends with the *Siahi Ki Bidhi*, which lists the recipe for the ink used throughout the scripture.

Notable studies by scholars such as Professor Sahib Singh (1996),[37] Gurinder Singh Maan (2001)[38] and Pashaura Singh (2000)[39] have identified multiple variations of historic Sri Guru Granth Sahibs. After visiting the Rylands, Pashaura Singh directly referred to the style of Punjabi MS 5 as a 'Banno Recension'.[40] Scholars have debated whether these passages formed the original text in 1604 and were subsequently removed, or whether they are 'spurious' passages which were inserted post finalisation. While these passages are not uncommon in historic Granths, analysis of the debates suggests the history is complex and often unclear. There is in places a disconnect between Sikh scholarly research and the general Sikh population. To some contemporary Sikhs the passages appear alien and raises concerns of interference and the tampering of the scripture. This has led some to subsequently challenge the legitimacy of recensions containing these passages and question the role they have in twenty-first century Sikhi. This required careful and extensive guidance.

Digitisation

Digitisation was led by senior photographer (now imaging manager) James Robinson, carried out using a Phase One IQ4 camera with 150-million-pixel resolution, capturing all elements of the manuscript including flyleaves, covers, edges and binding details. Over 940 high-resolution images were taken, providing a comprehensive digital object. In addition to the main text, many fragments from damaged, particularly loose end, folios were digitised. Each fragment is treated as a 'limb' of the Guru, adhering to the same respectful processes applied to the main manuscript.

Conservators Jasdip Singh and Jivanpal Singh from Pothi Seva[41] have expertise in working with historic Sikh and Punjabi manuscripts. Jasdip outlines the work they undertook for the project:

> It is important to note that, prior to and during the conservation process, a large number of fragments, some large and others minute, became completely detached from the Guru Granth Sahib. These fragments mostly belonged to the final section which is particularly important in helping to date and categorise Guru Granth Sahib manuscripts. The fragments ranged in size from large pieces with complete sentences, to partial words and finally a larger number of fragments with partial letters. In the case of Sikh Gurbani manuscripts there is a major spiritual importance attached to the form of the Guru. This makes it especially important to try and ensure all fragments are re-attached. The task of matching up the fragments was digitally outsourced to Pothi Seva. Pothi Seva used high resolution photographs of the recto and verso of each fragment and used photographic editing software to isolate the fragments from the background. These images were then overlaid onto digitised images of the manuscript. Pothi Seva scholars used a manuscript-based font to digitally 'complete' the areas of missing text and, through a process of trial and error, were able to correctly locate the original position of the majority of fragments. This was a collaborative effort which was only made possible through the high-quality digitisation of the manuscript.[42] (Figure 4)

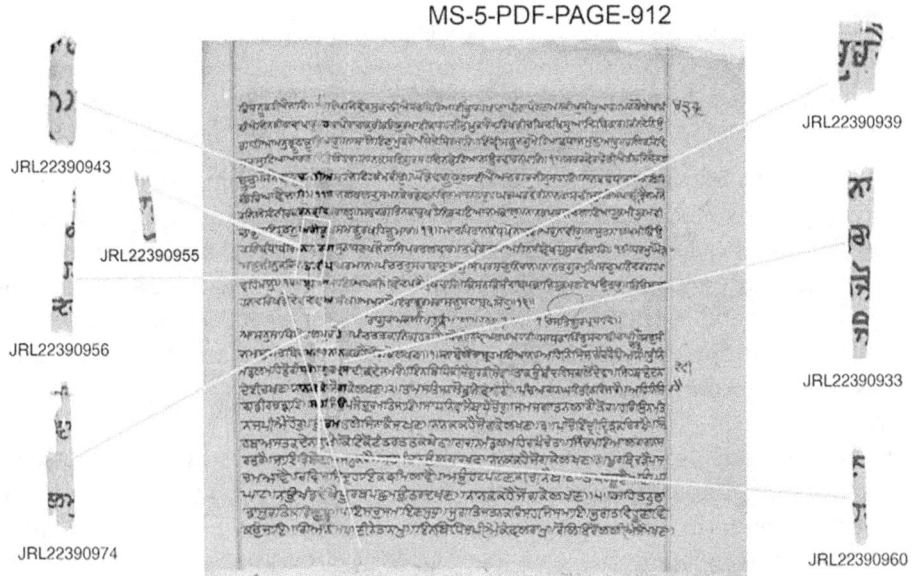

Figure 4 Digital reconstruction using digitised fragments and folio. Image provided by Jasdip Singh, Pothi Seva.

Recognising the Guru as the living embodiment of spiritual authority in Sikhi, the imaging team implemented a series of respectful practices during the digitisation process, matching the processes used in conservation:

- the Guru was transported carefully on a trolley and remained wrapped in a Rumalla during transit;
- shoes were removed and head coverings were worn both during transportation and while in the digitisation studio;
- a dedicated space was used exclusively, which was deep cleaned before and midway through the process;
- staff abstained from alcohol and tobacco for two days prior to, and during, the digitisation process;
- every effort was made to avoid turning one's back to the Guru during work;
- during breaks or pauses in digitisation, the Guru was carefully wrapped and respectfully stored.

While the utmost care is standard for all items that the imaging team digitise, the handling of the Guru was conducted at a slower pace with heightened mindfulness.

The digital version of the Guru is hosted on the library's digital platform, LUNA. It is freely accessible under a Creative Commons BY-NC-SA licence. However, the digital record includes clear guidance on the cultural sensitivities associated with it, written in collaboration with the community:

> The Guru Granth Sahib is considered by Sikhs to be the living embodiment of the Guru. Therefore, any printed version of the text is also regarded as an embodiment of the Guru and should be treated with the same respect. We respectfully request that you do not print images of the Guru Granth Sahib from this platform.

The digitisation of sacred or cultural objects risks becoming an extension of epistemological violence if not carried out without the meaningful involvement of the communities to whom the objects belong. Treating such items as mere artefacts, subject to digitisation without proper consultation, disregards their profound cultural and spiritual significance. This practice reflects a continuation of colonial assumptions, prioritising institutional authority and Western frameworks over the values and beliefs of the communities from which these objects originate.

Digitisation can also raise complex legal questions, particularly around intellectual property rights, copyright and the ownership of digital data. As Crawford and Jackson write in 'Stealing Culture, Digital Repatriation': 'Actions that have been historically referenced as "archaeological explorations" or "spoils of war" often provide fertile ground for considering the differences between ownership and possession. This critique becomes even more nuanced when the item is not simply a tangible object, but data.'[43] In 'Why Do We Digitise? The Case for Slow Digitisation', Prescott and Hughes state: 'Paradoxically, there is a risk that an emphasis on digitizing cultural treasures will undermine the claim that digitization opens up and democratizes access to cultural heritage. If digital libraries merely reiterate and reinforce long-standing cultural narratives and stereotypes, rather than enabling the exploration of forgotten and neglected collections, then they can become agents of cultural exclusion.'[44] Temi Odumosu adds, 'Digitization has come with critiques of power, bias, and legitimacy, since the institutional drive to reproduce the excessive scale of the colonial project as big data enacts its own forms of erasure.'[45]

The Sarr–Savoy report, officially known as the *Report on the Restitution of African Cultural Heritage*, is a 252-page document commissioned by the French government that aims to address the issue of restitution of African cultural artefacts acquired during the colonial period and currently held in French museums. One of the report's recommendations was to encourage full digitisation and open access to *all* material that was to be repatriated back to places of origin.[46] Odumosu draws out notions from the report: 'it has become clear that ethics of care requires more nuanced and holistic organizational mindsets to accommodate the vulnerabilities of postcolonial collections management'.[47]

In their response to the Sarr–Savoy report, Pavis and Wallace argue that 'the management of intellectual property is a cultural and curatorial prerogative, as the initial decision about whether and what materials to digitize. These prerogatives should belong to the communities of origin.' They argue clearly that intellectual property rights, open access and the digitisation of material must be discussed with and ultimately decided by the communities of origin. 'The current practice of Western governments and heritage institutions campaigning for and leading digitization projects according to Western values and priorities, such as open access, may be appropriate for their own cultural heritage. As applied to non-Western cultural heritage, it carries the potential to sustain the very colonial approaches the Report takes great care to denounce.'[48]

Overall, the Sarr–Savoy report sparked significant debate and discussion about the restitution of African cultural artefacts, both in France and internationally. The response by Pavis and Wallace was signed by 108 scholars and intellectual property law experts, as well as practitioners from cultural heritage institutions and organisations. While some progress has been made in implementing its recommendations, the issue remains complex and contentious, with ongoing efforts needed to address historical injustices and promote cultural equity.

The Guru in the Community

Initially, it was thought that access to the Guru could successfully be facilitated within the Rylands itself. However, an internal event quickly revealed that the library could not adequately accommodate demand and access while maintaining its regular public operations. Consequently, it was decided that a more suitable and meaningful location for future events would be a local Gurdwara, a space designed to host the Guru and deeply embedded in Sikh customs and traditions. During consultations with the community, many local Sikhs expressed their desire to see the Guru housed within a Gurdwara setting.

On Saturday, 13 April 2024, an event was held at the Sri Guru Gobind Singh Gurdwara in Whalley Range to coincide with Vaisakhi, one of the most significant celebrations in the Sikh calendar. For those involved in the project the event represented a major milestone, and the event was widely regarded as a success. The event received support from prominent national Sikh organisations, whose endorsements appeared on promotional materials, and was advertised on various media platforms in both English and Punjabi, the ancestral language of many Sikhs. The day's itinerary was thoughtfully curated by the community to align with Sikh customs and practices. The community took responsibility for transporting the Guru to the Gurdwara (Figure 5) and its safe return to the library, adhering to traditional customs rather than using university resources. Prominent Sikh scholars from across the country participated by reciting (Figure 6) and singing passages from the Guru, and the event garnered extensive media coverage including national broadcasts as well as international coverage on Sikh Channel.

An estimated 3,000 visitors attended, and feedback underscored the significance of the gathering being community-led. For many, this day symbolically marked the return of the Guru to the Sikh community for the first time since its contested acquisition in 1849.

Figure 5 The Sri Guru Granth Sahib ceremoniously leaving the library for the first time. © The John Rylands Research Institute and Library, University of Manchester.

Figure 6 The Sri Guru Granth Sahib being recited at the Vaisakhi celebration held at the Sri Guru Gobind Singh Gurdwara in 2024. This was the first time in perhaps 175 years the Guru had been consulted in this way, since its violent acquisition. © The John Rylands Research Institute and Library, University of Manchester.

Conclusion

Moving forward, institutions must recognise that digitisation is not a neutral act; it can be seen as a process shaped by the same epistemic structures that historically facilitated colonial exploitation. Without careful consideration and consultation, digitisation risks becoming another mechanism of epistemological violence, abstracting sacred objects from their cultural contexts and perpetuating a hierarchical framing of knowledge. Instead, it should serve as an opportunity to rethink and reframe institutional practices, emphasising collaboration, community involvement and the decolonisation of knowledge production. This requires not only technological resources but also a commitment to ethical engagement and a willingness to critically interrogate the power dynamics underlying museum and library collections and their interpretation.

This project is far from perfect or complete, but it represents an important step in challenging institutional barriers and building trust. By fostering community involvement, co-producing exhibitions and addressing the legacies of colonialism, we aim to create more inclusive and ethical cultural practices. However, it is essential to avoid complacency or self-congratulation. Decolonisation is not a final destination but an ongoing process that requires humility, accountability and a willingness to critically examine and transform institutional practices.

Dialogical curation, as demonstrated in this project, provides a valuable framework for fostering inclusive, responsive and ethically grounded approaches to cultural heritage. By prioritising collaboration and community voice, institutions can begin to address the epistemological violence of their histories and move towards more equitable and respectful stewardship of contested cultural artefacts.

Notes

1. Sri Guru Granth Sahib, hosted on LUNA, http://luna.manchester.ac.uk/luna/servlet/s/da0z5h [accessed 5 October 2024].
2. S. Manzoor-Khan, *Here, and Here and Here, We Have Always Been Here; Marginalised Histories Explored*, exhibition at the John Rylands Research Institute and Library, January–July 2024.
3. S. Manzoor-Khan, 'No Rylands Without Us', *Rylands Blog*, 18 January 2024, www.rylandscollections.com/2024/01/18/no-rylands-without-us-suhaiymah-manzoor-khan/ [accessed 5 December 2024].
4. G.C. Spivak, 'Can the Subaltern Speak?', in C. Nelson and L. Grossberg (eds), *Marxism and the Interpretation of Culture* (Urbana/Chicago: University of Illinois Press, 1988), pp. 271–313.
5. Ahmed Iqbal Ulla Race Relations Centre & Education Trust, 'South Asian Heritage Month 2021', www.racearchive.org.uk/south-asian-heritage-month-2021/ [accessed 15 November 2024].
6. Ahmed Iqbal Ullah Race Centre, www.library.manchester.ac.uk/aiu-race-centre [accessed 13 January 2025].

7 Ahmed Iqbal Ulla Race Relations Centre & Education Trust, www.racearchive.org.uk [accessed 13 January 2025].
8 Message from John Hodgson to Gurtek Singh, 13 October 2024.
9 A Sanskrit term which translates to 'enlightener'.
10 Nihung Santhia, 'History of Pehla Parkash Sri Guru Granth Sahib', www.nihungsanthia.com/post/history-of-pehla-parkash-sri-guru-granth-sahib [accessed 10 November 2024].
11 M. A. Macauliffe, *The Sikh Religion: Its Gurus, Sacred Writings and Authors* (Oxford: Oxford University Press, Volume 1, 1909), p.xvi.
12 Basics of Sikhi, 'Guru Granth Sahib Ji', www.basicsofsikhi.com/post/guru-granth-sahib-ji [accessed 10 November 2024].
13 J. Hodgson, 'Spoils of Many a Distant Land: The Earls of Crawford and the Collecting of Oriental Manuscripts in the Nineteenth Century', *The Journal of Imperial and Commonwealth History*, 48:6 (2020), 1011–47.
14 L. Smith, 'Collecting and Colonialism', *The John Rylands Library Medium*, 21 June 2021, www.medium.com/special-collections/collecting-and-colonialism-c6ffa8323c96 [accessed 5 December 2024].
15 National Library of Scotland, Acc. 9769, Crawford Library Letters, vol. 9, f. 143, Letter from Bernard Quaritch to Lord Crawford, 26 June 1862.
16 'Handlist of Hindustani, Marathi & Punjabi Manuscripts in the John Rylands Library' (unpublished catalogue, John Rylands Library, n.d.).
17 National Library of Scotland, Acc. 9769, Crawford Library Letters, vol. 9, f. 144, Letter from Bernard Quaritch to Lord Crawford, n.d., 1862.
18 J. Hodgson, 'The Rylands' Sri Guru Granth Sahib: From Conquest to Co-Curation', at Research Libraries UK Special Collections and Heritage Network, 7 October 2022.
19 S. Vawda, 'Museums and the Epistemology of Injustice: From Colonialism to Decoloniality', *Museum International*, 71:1–2 (2019), 72–9.
20 *Ibid.*
21 Y. G. Woldeyes, '"Holding Living Bodies in Graveyards": The Violence of Keeping Ethiopian Manuscripts in Western Institutions', *Media / Culture Journal*, 23: 2 (2020).
22 P. Singh, *The Guru Granth Sahib: Canon, Meaning and Authority* (Delhi, 2003).
23 M. Furness, *Panjabi MS 5 Condition Report* (Manchester: John Rylands Library, 2010).
24 Greater Manchester Combined Authority, *Census 2021 Briefing: Religion*, www.greatermanchester-ca.gov.uk/media/8081/census-2021-briefing_religion_final_so.docx [accessed 15 November 2024].
25 Office of National Statistics, *2001 Census Key Statistics*, www.nomisweb.co.uk/census/2001/uv015 [accessed 15 November 2024].
26 The Sikh temple, the natural home of the Guru and important community hubs.
27 Sri Dasam Granth is sometimes referred to as Sri Dasam Guru Granth Sahib. It is a collection of compositions attributed to the Tenth Guru, Sri Guru Gobind Singh. Punjabi MS 6 is hosted on LUNA, luna.manchester.ac.uk/luna/servlet/s/shin65 [accessed 15 November 2024].
28 A Pothi is a historic manuscript that contain the hymns and spiritual teachings of Sikh Gurus and other saints. They were traditionally a condensed version of the Sri Guru

29 *The Golden Temple of Amritsar*, www.goldentempleamritsar.org [accessed 10 November 2024].
30 L. Unruh, 'Dialogical Curating: Towards Aboriginal Self-Representation in Museums', *Curator: The Museum Journal*, 58 (2015), 77–89.
31 *Ibid.*
32 Punjabi MS 1 is hosted on LUNA, luna.manchester.ac.uk/luna/servlet/s/1ldaf7 [accessed 15 November 2024].
33 Punjabi MS 3 is hosted on LUNA, luna.manchester.ac.uk/luna/servlet/s/q15g1o [accessed 15 November 2024].
34 Handwriting analysis of the Chalitar, which details the dates of when each Guru left the world, indicates a change in the scribe for the insertion of the Tenth Guru. This suggests it may not have been part of the original text but a later addition.
35 *Rattanmala Mahala Pehla* translates as Garland of Jewels and is attributed to Guru Nanak Dev Ji, although scholars dispute this.
36 *Haqiqat Rah Muqam Raje Shivnabh Ki* is not attributed to any writer. The passage relates to a meeting between Guru Nanak Dev Ji and Raja Shivnabh in Sri Lanka.
37 S. Singh, *About Compilation of Sri Guru Granth Sahib*, trans. S. Dalip Singh (Amritsar: Kulwant Singh Suri, 1996).
38 S. M. Gurinder, *The Making of the Sikh Scripture* (Oxford: Oxford University Press, 2001).
39 Singh, *The Guru Granth Sahib*.
40 *Ibid.*, p.220.
41 Pothi Seva, www.pothiseva.net [accessed 15 November 2024].
42 Email from Jasdip Singh to Gurtek Singh, 28 January 2025.
43 N. Crawford and D. Jackson, 'Stealing Culture: Digital Repatriation (A Case Study)', *University Museums and Collections Journal*, 12:2 (2020), 77–83.
44 L. Hughes and A. Prescott, 'Why Do We Digitize? The Case for Slow Digitization', *Archive Journal*, first published online September 2018, www.archivejournal.net/essays/why-do-we-digitize-the-case-for-slow-digitization/ [accessed 5 December 2024].
45 T. Odumosu, The Crying Child: On Colonial Archives, Digitization, and Ethics of Care in the Cultural Commons, *Current Anthropology*, 61:22 (2020), 209–302.
46 F. Sarr and B. Savoy, *The Restitution of African Cultural Heritage: Toward a New Relational Ethics* (Paris: Ministère de la culture, 2018).
47 Odumosu, 'The Crying Child', 209–302.
48 M. Pavis and A. Wallace, 'Response to the 2018 Sarr–Savoy Report: Statement on Intellectual Property Rights and Open Access Relevant to the Digitization and Restitution of African Cultural Heritage and Associated Materials', *Journal of Intellectual Property, Information Technology and E-Commerce Law*, 10:2 (2019), 115–29.

Imaging the Gaster Jewish Amulets in the John Rylands Research Institute and Library

PHILIP ALEXANDER, UNIVERSITY OF MANCHESTER
JAMES ROBINSON, UNIVERSITY OF MANCHESTER
ELIZABETH EVANS, UNIVERSITY OF MANCHESTER
AMIN GARBOUT, UNIVERSITY OF MANCHESTER
JO CASTLE, UNIVERSITY OF MANCHESTER
TONY RICHARDS, UNIVERSITY OF MANCHESTER
IRA RABIN, BAM, BERLIN

Abstract

This article surveys the advantages of using digital imaging to enhance the catalogue of a collection of Jewish amulets in the Rylands Library Manchester. Like manuscripts, there are obvious advantages in having images of amulets online, in terms of making them more widely accessible to scholars, in allowing more easy comparison one with the other, in promoting conservation by reducing handling, in aiding detailed analysis (e.g. through magnification) and in encouraging public engagement; but amulets pose their own distinctive problems: they are often 3D, which makes photographing them more difficult than photographing the flat surface of a manuscript. The article describes how 3D photogrammetry was used to solve this problem. It also describes how X-ray computed tomography was used to discover what was contained in a number of sealed amulets, without damaging the artefacts. These techniques are costly: the article concludes by calling for the development of criteria to guide us when they should be applied.

Keywords: Rylands Library; amulets; 3D photogrammetry; X-ray computer tomography

The purpose of this article is to record the experience – the achievements, the failures and the pitfalls – of using digital imaging to facilitate and enhance a project to catalogue the Jewish amulets in the John Rylands Research Institute and Library, University of Manchester.[1] There are about 270 of these amulets, all of which belonged at one time to the Jewish savant and collector Moses Gaster (1856–1939). Gaster was a polymath who, although not a wealthy man, amassed a remarkable collection of Jewish and Samaritan manuscripts and artefacts, the bulk of which entered the John Rylands Library in the 1950s either by purchase from the Gaster family, donation or loan. The transfer was facilitated by the mediation and generosity of the local Jewish community, in which Gaster was held in high esteem. His eldest daughter, Phina Emily, had married Neville Laski of the Laski family of Manchester, which was prominent in Jewish communal and civic life in the city. The amulets formed part of the wider Gaster collection of manuscripts and artefacts.[2]

The cataloguing of the amulets constituted the third phase of a project to catalogue the complete Gaster collection at the John Rylands Library. Two earlier phases of this had been more or less completed. The first phase involved cataloguing around 226 codices (complete and incomplete) in Hebrew script in the library; these included not only the Gaster manuscripts but also forty-one codices already in the library, which had belonged to the Crawford and Spencer Collections, among which were magnificent illuminated manuscripts, such as the famous thirteenth-century Rylands Passover Haggadah (Hebrew MS 6). Images of all these manuscripts – along with accompanying catalogue records – are now available online in Manchester Digital Collections (MDC), and have been uploaded to the Jewish National and University Library's KTIV database, which aims to make available online all the manuscripts in Hebrew script scattered throughout the world.[3] The second phase, begun while the first was in progress, involved imaging and cataloguing the nearly 15,000 manuscript fragments acquired by Gaster which had originated in the Genizah (a storeroom for discarded manuscripts) in the Ben Ezra synagogue in Fustat (Old Cairo). These two phases are now complete. The images and accompanying catalogue records of the Genizah fragments are available in the Library Digital Collections (Luna), and have been uploaded to the Friedberg database, which is reassembling – as far as is possible – in cyberspace the contents of the Cairo Genizah, now held in discrete collections throughout the world.[4] The third phase involved cataloguing the Jewish amulets. This is also now complete, and images and records of these are available on MDC.[5]

The Advantages of Imaging

Imaging manuscripts and putting the images online offers a number of well-known advantages. The first is accessibility, since digital images allow scholars, researchers and the general public to study the artefacts remotely. This was crucial in the case of the Gaster codices, which are very diverse in content and language. Almost all the texts are in Hebrew script, but the *languages* written in that script are quite different: not only Hebrew and Aramaic, which since antiquity had used the 'Hebrew' alphabet, but also Judeo-Arabic, Judeo-Persian, Judeo-Tatar (Karaim), Judeo-Occitan (Judeo-Provençal), Ladino and Yiddish, among others. The texts were often fragmentary, and sometimes contained hitherto unknown works. All this diversity taxed the in-house expertise of those making identifications and drawing up the catalogue records. With the digital images available we were able to circulate the problematic texts to known experts in specific fields and benefit from their opinions. Here scholarly networks such as the Hebrew Codicology and Palaeography Facebook group proved invaluable.

Circulating a text which puzzled us through these channels usually elicited helpful comments or identifications, but this was only possible when we had high-quality digital images to share.

The second advantage is comparability: having large numbers of digital images of manuscripts available online allows them to be compared more easily. This comparison is important not only for text-critical purposes (collating variant readings of the same text) but for palaeography (the handwriting of texts), codicology (how the texts are formatted and put together) and iconography (the comparison of the imagery and aesthetics used in illustrated texts). Large databases of images also lay the foundation for rejoining fragments of the same manuscript now held in different repositories. This is potentially important for the Cairo Genizah texts; many of them are fragmentary, and scattered across different libraries. In the past scholars who happened to see two fragments might spot that they make a join, but one possibility that is now emerging is that AI may be trained to work through a large database of digital images of Genizah fragments to discover matching edges.

The third advantage of imaging is conservation. Many of the artefacts are fragile and in a poor physical condition. Many are made of organic materials such as parchment, paper or even earth; some are made from delicate fabrics or have intricate and moving parts. All have deteriorated over time. High-resolution photography ensures that a record is preserved for posterity, and at the same time reduces the need for direct physical handling, and so contributes towards conservation. While some research purposes will still necessitate physical inspection of artefacts (there are ways in which images – however good – can be misleading, such as in relation to colour and size), most research can be carried out satisfactorily on the basis of images. This also reduces the need for travel, incidentally producing an ecological bonus.

The fourth advantage is analysis. High-resolution digital imaging has proved a powerful tool in analysing the artefacts, because of the possibility it offers to manipulate the images (e.g. by magnifying portions of the image, changing the contrast or filling in lacunae by cutting and pasting words or letters from elsewhere in the text). Fine details can be studied, faint writing read and corrections untangled. Digital images can do everything that used to be done by examining the physical object through a magnifying lens, and much more besides. Surface damage can be seen through: manuscripts can be photographed in different wavelengths of light and through different filters, thus revealing features that are invisible or illegible to the naked eye even with magnification. Photographic techniques have been developed to read palimpsests and to digitally 'unroll' and read damaged scrolls, which could not be physically unrolled without further damaging or destroying them.

The fifth and final advantage is public engagement. Online images of cultural artefacts are not just aimed at scholars but have for some time been playing an increasingly important role in making heritage available to a wider audience. The Gaster Project at Manchester is an important outreach by the University of Manchester to the city's sizeable Jewish community (which, as noted above, originally made possible the acquisition of the collection by the John Rylands Library). It has therefore helped mediate Jewish heritage to the community that culturally owns it. This has been reinforced by regular blogs, an online exhibition and talks to community groups, drawing attention to the collection.[6] This has helped the university and the library fulfil their public remit, and has gained goodwill and support for what could appear from the outside to be remote 'ivory tower' research, of little interest or relevance to people's everyday lives. None of this would be feasible without digital imaging.

The Challenge of the Amulets

All these advantages applied in the case of the Gaster Jewish Amulets, but the amulets presented their own distinctive challenges compared with the manuscripts that had been the focus of the first two phases of the project. This led us to set them aside as a separate sub-project.

One of these challenges was how to definite an 'amulet'. Our collection of 'amulets' embraces a very diverse range of artefacts, some with extensive written texts, others without any writing whatsoever. Some are on metal, others on parchment or paper; some are worn on the person, others are hung on a wall or located at significant points in the house, such as the door. Most display symbols of various kinds. What links these diverse objects is not material, form or precise content, but social function: all are designed to use recognised magical means to protect their user from harm. The cataloguing of Hebrew-script manuscripts has evolved over the previous two centuries, and there is now a consensus as to which elements are sufficiently important to be included in a catalogue: these include date, authorship (if appropriate), provenance and language(s). A number of these fields are appropriate also to amulets, but there are features of amulets that may be significant – such as their shape, material and colour – which are rarely, if ever, significant for manuscripts. It is unclear exactly which elements of amulets *are* significant, and the significance of some features may vary from object to object or from context to context. Which are purely aesthetic and which magically functional? The distinction between amulets and jewellery has always been blurred.

An amulet can be seen fundamentally as any object intended to protect someone against demonic harm by virtue of some magical property that it contains. That apotropaic magical power may reside in a written magical formula or formulae, or it may inhere in the shape, the material or the colour of the amulet, or the symbols it displays, without any written text, or in all of these. There is as yet no consensus as to what the descriptive fields should be for a catalogue entry for an amulet. The Victoria and Albert Museum in London has created online a database of images of 186 amulets in its collections, for example, but the accompanying records are very basic; they focus on the date and provenance of the artefact and on its description, but they say nothing about its possible use or its distinctive magical features.[7]

The use of amulets is widespread among Jews. Jewish amulets share many features with the amulets of other communities (e.g. Christian, Islamic and pagan);

magic is notoriously eclectic and syncretic, and it is sometimes impossible to assign a given amulet to a particular tradition. The most distinctive marks of Jewish amulets are the use of the Hebrew script (for texts either in Hebrew or in Jewish Aramaic, which are often verses from the Hebrew Bible), the use of various Jewish names of God (above all the Tetragrammaton, YHWH), the invocation of Jewish angels (e.g. Gabriel, Michael, Raphael), the execration of Jewish demons (e.g. Lilith, Agrat bat Mahlat) and the use of Jewish symbols such as the hexagram (the so-called Star of David, or Magen David). Amulets are still widely worn, and some continue to believe in their efficacy. For others they are little more than jewellery, although (like the Star of David) they may additionally serve as a marker of Jewishness, in the same way that a cross does for Christians.[8]

The sheer diversity as well as the physical condition of many of the items in the collection also proved challenging for the photographers. Each individual item required specialist handling, support and different photographic techniques. The requirements of photographing a medallion (such as GA 55a) are very different from those required to photograph a necklace (GA 4 and 5), an ancient spearhead (GA 16 and 57), a prayer-shawl (GA 68), a wimpel (Torah binder) (GA 14a) or a bonnet worn by a baby boy during circumcision (GA 54). Many pieces in the collection, such as necklaces, bracelets and anklets, comprised many parts. Each had to be arranged in the frame to show the maximum amount of detail without losing the sense of the piece as a whole. Most of the objects were 3D in a way that a manuscript page is not, and needed to be viewed from all angles. They could therefore not be adequately captured by a static 2D image.

The amulets in the collection also vary in condition and fragility. All items required the wearing of nitrile gloves, and some required delicate handling and positioning using a variety of tweezers, soft brushes and acrylic spatulas. One of the more difficult items to image was GA 39, which incorporated the head of a great stag beetle, with antennae and moving mandibles still attached. GA 27 is a solid, cone-shaped, compacted lump of earth with a leather strap embedded in it. It is very fragile because the earth has dried out, particularly around the strap. The label on the amulet reads: 'Against fever. Made of earth taken from the precincts of the Holy Mosque at Mecca. Dipped in blood of one of the sacrifices brought there'. Although we have not yet analysed the blood, it is probably sheep or goat. Since it was difficult to do justice to this object with static images, we included a video that allowed the object to be rotated and manipulated.

The Photography: Lighting

High-end photographic equipment was used to handle these challenges. All imaging was carried out using a Phase One IQ4 digital back, XF body and 120-mm Schneider Kreuznach macro lens, also utilising 2x extension tubes for even closer work. Various lighting techniques were used to minimise reflection and distracting highlights, and to enhance the visibility of inscriptions and surface texture and detail. These included:

- *Raking light*, i.e. using a single light source rather than multiple sources to enhance surface detail and texture.
- *Focus stacking*. Particularly with macrophotography, depth of field is incredibly shallow, meaning only a very small area of the field of the image is in focus. Stacking allows for multiple images to be taken at different levels of focus, then merged together using software, producing a single image that shows the item in complete focus.
- *Transmissive light table*. Using a transmissive light table, images can be taken of objects with no distracting shadows. Photographing transparent and semi-transparent objects in this way can show structure, internal detail and clear shape. The table consists of a tempered glass sheet across four supports and a base and is used for imaging papyrus and photographic materials such as negatives and transparencies.

Multiple images were captured for each item, to make sure that all views of each piece were included. For more complicated items, digital images were captured of key areas of interest.

3D Photogrammetry

3D photogrammetry was used as a way of tackling the challenge of the three-dimensionality of the objects. This technique allows for a 3D model to be created of an object, which can be rotated and interrogated virtually. Multiple images are captured of the item from all angles, often utilising hundreds of positions to create an accurate model. For the amulets, the image was captured using a Sony A7rIV, Bluetooth-controlled turntable and light tent. The Bluetooth-controlled system automates the rotation and capture of the images in sequence. Images are then fed into software that recognises each individual position of the image as though the camera had moved around the object. Using these anchor points in each image, the software can then build the 3D model. Texture, colour and shading are also recorded from the images and laid over the model. Once complete, the 3D objects are hosted on Sketchfab, which allows for the embedding of the models into the MDC collection alongside the standard images and Text Encoding Initiative (TEI) metadata.

Images produced by photogrammetry were included as appropriate alongside standard images in the MDC database. This was important since the standard images displayed important technical information necessary for the correct reading of the image. Correct colour balancing is imperative in imaging, and each image contains a Golden Thread colour target, as well as inch and millimetre scales and a dots per inch (DPI) resolution chart. This information is not displayed in the photogrammetry image, which may, therefore, give a misleading impression of size and actual colour if viewed on its own. This information could be added later digitally, but would be an estimation, not a scientific measurement.

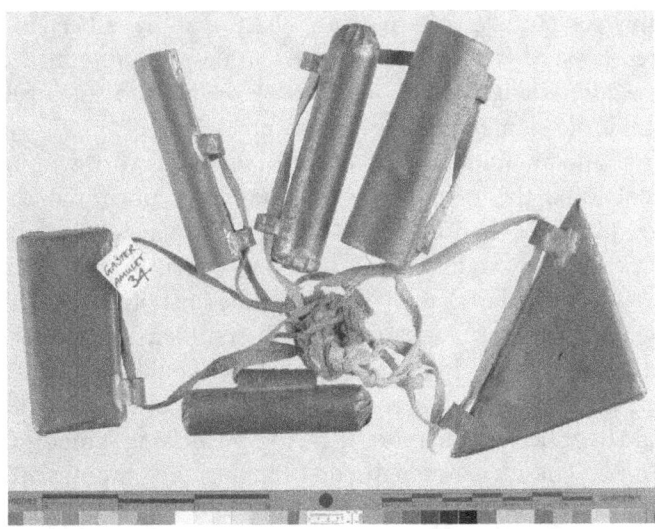

Figure 1 John Rylands Research Institute and Library, Gaster Amulet Collection, Gaster Amulet 34. Image © The University of Manchester.

Sealed Amulets: Gaster Amulet 34

There was an added complication. Some amulets are sealed containers, with (probably) a text written on paper or parchment inside. A cluster of six of these is numbered GA 34 in the MDC database (Figure 1). GA 34 comprises four metal tubes (34b, 34d, 34e and 34f) of slightly different size. Their length ranges from 65–85 mm, and their diameter from 15–25 mm. There is also one thin rectangular metal box (34a: 65 × 30 mm) and one thin triangular metal box (34c: 90 mm along its longest side, and 65 mm along the other two sides). The boxes were made by beating over a narrow flange along all the edges of the two plates and then sealing the flanges together with solder. The tubes were sealed in two ways: either short cuts were made in the end of the tube and then the strips thus created were beaten over to close the end and sealed with solder, or else a round, flanged metal disc matching the inner diameter of the tube was inserted into the end of the tube and soldered in.

Each container had a small loop or loops soldered to it, made of the same material, which allowed it to be hung on a lanyard around the neck. This is standard practice, since it is generally necessary, if the amulet is to be personally effective, for it to be brought in close proximity to the body of the person whom it is meant to protect: to be worn in some way. Two of these lanyards, made of fabric – one red and one white – are preserved. The catalogue entry suggests that all the containers were all looped on to *one* lanyard and that this formed a single necklace, but this is unlikely, since the result would have been very clunky: all the containers would have slid down higgledy-piggledy to the bottom of the lanyard. Each one may originally have been a pendant on its own lanyard, and they may have been strung together simply for storage. Alternatively, there may have been *two* necklaces, one on the red lanyard (rectangular box + 2 tubes one on each side of it), the other on the white (triangular box + 2 tubes one on each side of it). Necklaces made up of multiple elements are well known.[9]

The material and manufacture of the containers is worth noting. It is very simple and functional; the metal is base and looks the same for all the containers. We concluded in-house that, because it attracted a magnet and shows signs of corrosion, it probably contains iron; it was subsequently scientifically analysed by Ira Rabin and Mark Furness of BAM (Bundesanstalt für Materialforschung und -prüfung, Berlin), who concluded: 'XRF identified iron as the basis metal of the amulet. So we assume it is made of steel. We could also identify the solder materials: lead and tin.'[10] There is a marked lack of decoration. With a little effort the boxes could have been made more attractive by incising symbols on them such as a Magen David. These are not high-end products: there was no attempt to turn them into jewellery. Yet the containers are neatly and skilfully made, and involved collaboration between an amulet-writer and a tinsmith.

That they belong to the low end of the market (like much of the Gaster Amulet Collection) is confirmed by comparing them with similar amulets. Three of these are featured in the catalogue for the exhibition *Angels and Demons: Jewish Magic through the Ages*.[11] In two of these, the cases were made of silver, and in one example of 22 kt gold. The gold container is attached to a gold chain, while one of the silver amulets was decoratively incised with Hebrew letters to enhance its visual impact. It is clear, at least in the case of the gold container, that one end has since come off, enabling the scroll inside to be removed. Another example is featured in the catalogue for another exhibition, *Magie: anges et démons dans la tradition juive*.[12] Here the case is of chased silver, and the parchment scroll has been extracted and unrolled beside it; it contains a carefully executed Qabbalistic symbol, the Sefirotic Tree. These examples also illustrate how persistent this type of amulet was over time and space: one is from Algeria and dates to the end of nineteenth century; another is from Iran and dates to *c.* 1930; a third is from Iraq and dates to *c.* 1920; and a fourth is from Israel (El-Jish) and dates to the late fifth century CE. Gaster 34 is probably North African and dates to the early twentieth century. The ultimate model for this type of amulet is the Mezuzah: a container attached to the side-post of a door, containing passages from the Torah written on parchment. The idea of binding the spell to the body in order to make it personally effective may owe something to the analogy of tefillin (phylacteries): leather boxes containing passages from the Torah, which are strapped to the arm and head at times of prayer.[13]

Based on similar examples, we could predict that the containers housed magical spells written on parchment or papyri: protective passages from the Bible and similar authoritative texts, along with angelic and divine names and magical symbols. These formulae would have been copied out of magical recipe books and tailored to meet the needs of the client. The challenge was discovering what was inside the containers without opening them. Damaging the artefact in this way would be contrary to current curatorial practice unless there was some overriding consideration to the contrary (such as was not evident in this case). Broadly speaking, we were confident as to what we would find if we opened them; but until we could read their contents, we could not be certain even that this is a *Jewish* amulet, since there are no external signs identifying it as such. Sealed amulets were used also by non-Jews, and there are non-Jewish amulets in the Gaster collection. Nor was it beyond the realm of possibility that the spells might turn out to be pure hocus-pocus: unscrupulous amulet-writers were not above palming off on clients supposed formulae, which had no authenticity whatsoever. If they were sealed in a container, the client simply had to take the amulet-writer's word for it. The text might, however, be genuine and even contain useful information such as the client's name, the amulet's provenance and the specific condition that it was meant to protect against.

X-ray Computed Tomography

This was the state of affairs when we completed the entry for GA 34 in the online database, but the possibility then arose of seeing inside the containers without physically damaging them through the use of X-ray computed tomography (CT). The X-ray CT imaging was provided by the University of Manchester's Henry Moseley X-ray Imaging Facility, part of the National Centre for X-ray Computed Tomography (NXCT). This is a UK national research facility with centres at University College London and the universities of Warwick, Southampton and Manchester. It offers imaging for a wide variety of academic disciplines as well as commercial industries, and costs around £800 to £1,500 per day for X-ray CT scanning. Being part of the University of Manchester allows the Rylands Library to book five days of free 'beam time' a year, funded by the Engineering and Physical Sciences Research Council (EPSRC) as part of a scheme to support research, teaching and public engagement.

No special collections material from the Rylands had ever been imaged in such a way, and the amulets were selected as a perfect candidate for trials, especially GA 34, which contained sealed objects. This method of interrogating an object is a known means of deepening our understanding of the material of certain heritage artefacts. Not only can it help to identify and distinguish between distinct materials that may not be visible to the naked eye, it can also give insights into the construction of objects and detect areas of internal damage. Above all, as in the case of GA 34, it can detect what is hidden inside sealed containers. It must be stressed that this was a pilot exercise, aimed at identifying the sorts of things that X-ray CT can do, with a view to applying the technology to other artefacts in the future; it involved a steep learning curve for all concerned – photographers, cataloguers, curators and conservators – since none of them had been involved before in such an exercise.[14]

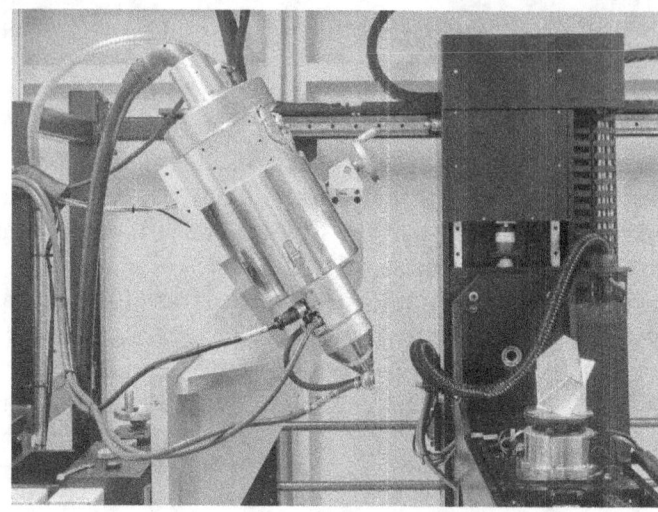

Figure 2 John Rylands Research Institute and Library, Gaster Amulet Collection, Gaster Amulet 34a inside its case in the High Flux Bay. Image © The University of Manchester.

Safe transportation of the amulets off site to the NXCT facility was arranged with Collection Care. Each of the metal casings that form GA 34 was separated and housed individually in custom-made archival boxes with bespoke foam inserts. These boxes not only protected the amulets during storage and transportation, but also provided the perfect mount for imaging, such that the amulets did not need to be removed from the housing during the scanning process (Figures 2 and 3).

All the containers of GA 34 were scanned. Low-resolution CT data was collected for 34a, b, c and d. High resolution CT data was captured for 34b only. Additional scans failed due to a technical issue with the scanner, a reminder of how chancy this process can be. The scans confirmed that there were indeed texts on paper/parchment inside 34a and 34b. In the case of 34b, this took the form of a roll that fitted tightly into the tube. Judging by the number of turns of the roll and the diameter of the tube – which it completely filled – it would be possible to roughly calculate the length of the unrolled scroll, which would be quite long. It would be something like the scroll shown in Bohak and Hoog, *Magie: anges et demons*, p. 113 (cat. 179), which fitted into the silver container lying beside it.

The image of the text inside 34a was somewhat surprising. Given the rectangular shape of the case, one might have expected a folded piece of parchment/paper, but the images that emerged clearly showed that it was first loosely rolled up and then squashed flat to fit into the box (Figures 4 and 5). We did not manage to achieve a successful scan of the inside of the triangular case (34c), but it is hard to see how a rolled text could have been fitted into the shape of the box; it is more likely that it

Figure 3 John Rylands Research Institute and Library, Gaster Amulet Collection, Gaster Amulet 34b in the smaller Heliscan MicroCT scanner. Image © The University of Manchester.

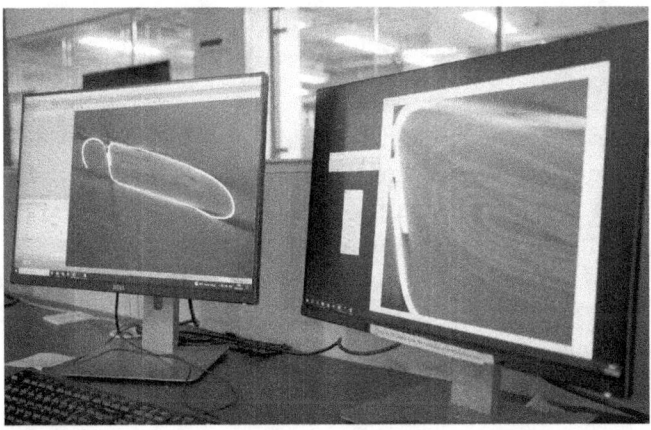

Figure 4 Cross-section of Gaster Amulet 34a end-on. Image © The University of Manchester.

was folded into a triangular shape. The triangle seems to have been a significant shape in Jewish magic, and it is reproduced in amulets in various ways. How the folding may have been achieved is illustrated by GA 83: this long, thin, parchment strip was originally folded into a triangle and is still stored in that way. It may

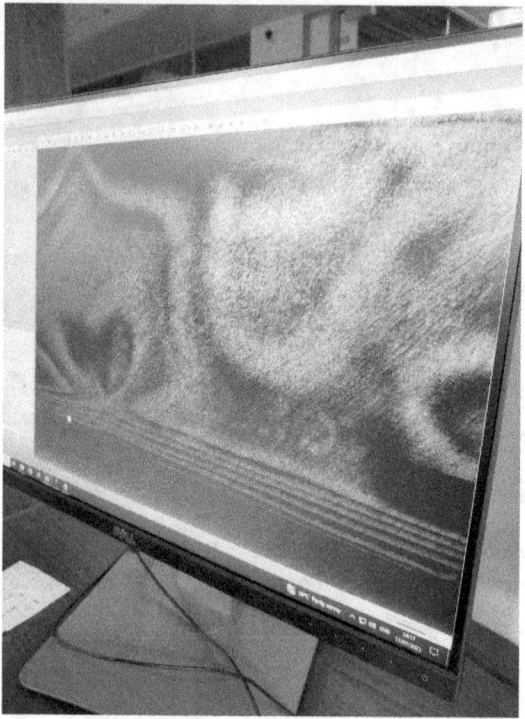

Figure 5 Cross-section of Gaster Amulet 34a from the top; note the faint traces of Hebrew letters. Image © The University of Manchester.

well have been originally housed in a triangular case such as GA 34c. Many parchment strips in amulet collections are without their containers; it seems likely that amulet collectors regularly opened the containers to retrieve the scrolls and then discarded the containers unless these had some intrinsic value.

The data was reorientated and digitally manipulated in Thermofisher's Avizo software, so that views parallel to the paper's surface could be seen. The lettering on the scrolls could be seen as a contrast between the ink and the paper/parchment, which was manually labelled in the software and rendered (Figure 6). This distinctly resembled Hebrew square script, and some of the letters looked incomplete. It is unclear why these particular letters emerged, but it may have had something to do with the thickness of the ink, since the Hebrew letters were probably shaded (that is to say, the various strokes that made them up were of various thicknesses of ink depending on how hard the scribe pressed down, and hence the amount of ink discharged from the pen). Each letter was made up of strokes of varying thickness, and only the thicker strokes were picked up in the scan. In some cases, the letters appear reversed or overlaid, probably the result of the folding or rolling of the paper/parchment.

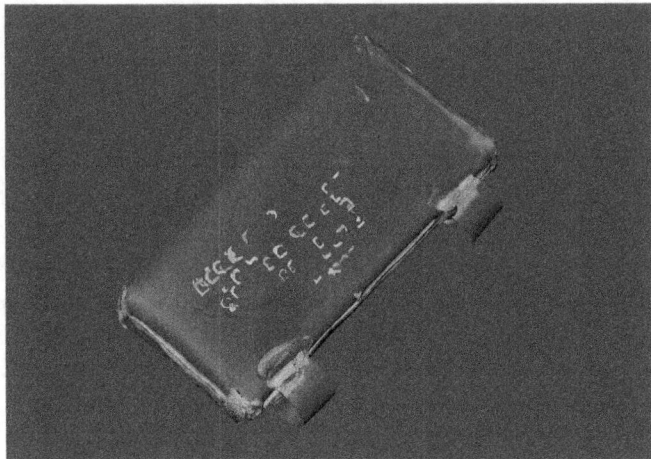

Figure 6 Traces of square Hebrew letters in Gaster Amulet 34a. Image © The University of Manchester.

Reflections on the Scanning Process

The scanning process proved instructive for all concerned. It demonstrated the value of CT scanning in certain circumstances, since we were able to show that at least two of the containers did contain written texts and that – at least in one instance – the writing was in Hebrew script, confirming that it was a Jewish amulet. The question naturally arises as to whether further processing of the data will reveal more of the writing and whether another scan with different parameters might yield better results.[15] Would the time, effort and cost be worth it? Cost should never be ignored, and scanning consumes scarce resources in terms of time, expertise and kit. At present, X-ray tomography can only be carried out in specialist facilities due to the high energy X-rays generated and the need for substantial radiation shielding for operators; it is therefore unlikely to become widely and inexpensively available in the foreseeable future. Nor does the fact that a procedure can be done mean that it should be. The expenditure of time and resources must be justified in terms of the cultural/historical significance of the objects scanned and the potential importance of the results that might be obtained. Fundamentally, it is the role of curators and scholars to determine the 'value' of the object, but that is not always straightforward.

As we noted in passing, the Gaster Amulet Collection represents – on the whole – the 'low' end of the market: the sorts of amulets that the poor could afford. Other collections tend to represent the 'high' end, as is attested by the use of precious metals such as silver and gold, and their production by highly skilled craftsmen. Paradoxically, this gives a certain cultural value to the Gaster Collection, since low-end amulets rarely survive. The rate of destruction of amulets is very high;

many were personalised and were simply thrown away when their owners died. It is hard to find a Jewish amulet that dates to before the early nineteenth century.[16] Those that were kept were often the ones that look valuable and are skilfully made, rendering them aesthetically pleasing as well as functional. Unsurprisingly, these are the amulets that catch the eye of modern collectors and end up in museums.

Cultural value should not be confused with material value. Nevertheless, the low value of GA 34 is determined by the fact that we have many comparable examples and that the written texts inside some of them have been retrieved. In the case of GA 34, we had no expectation of revealing anything substantially new. This process proved invaluable as a training exercise and a way of upskilling those involved, however. CT scanning and other advanced imaging techniques are an important tool for the study of cultural artefacts, but they will only be selectively available; there therefore needs to be a rational policy for selecting the objects to which they should be applied.[17]

Moreover, in the cost–benefit analysis, it is always important to assess the potential damage to the artefact. In the case of GA 34, the scan provided a non-intrusive means of discovering what the cases contained without having to open them, and so conformed to best curatorial practice. Nevertheless, the handling, transportation and scanning (e.g. subjecting the paper/parchment to X-rays) still needed to be assessed from a conservation perspective. Here the utmost care was taken not to degrade the artefacts, but that involved a monetary cost.

Another lesson learned from the process was the need for the curators and scholars to understand the technicalities involved and, conversely, for the technicians to understand what curators and scholars are trying to do. This learning curve can be steep, particularly on the side of the curators and scholars. They need to able to give the clearest possible brief to the technicians and interpret accurately the scans or photographs that are presented to them. Imaging is the future, but it is presently in its infancy. The use of AI to aid and streamline analysis is likely to transform the volume of data that can be analysed; machine-learning models have already enabled faster and more precise processing of X-ray CT data in a number of fields. This means that training in these technical processes needs to be added to the training of those who work with heritage artefacts. The baseline for interpretation, of course, remains the observation of the object with the naked eye. This was brought home in the present case by the fact that a video mp4 of the rectangular box (GA 34a) showed it as gold in colour when in reality it is a dull grey. It turned out that this colour was arbitrarily assigned by the technician. The same video also distorted the thickness of the solder compared with the naked eye, and exaggerated the depth of a small area of damage on the top right-hand corner of the case, close to one of the suspension loops.

Despite limitations due to the nature of the CT scan data, CT scanning provides invaluable information that can supplement the analysis of the object through observation by the naked eye, as well as by surface-imaging techniques. It is an important new tool in the researcher's toolkit.

Acknowledgements

This work was supported by the National Research Facility for Lab X-ray CT (NXCT) through EPSRC grant EP/T02593X/1. The figures in the text above are taken from J. W. Robinson, 'X-Ray Tomography of Gaster Amulet 34', posted on the *Rylands Blog* on 28 September 2023 (rylandscollections.com/?s=X-Ray+Tomography).

Notes

1. The collection is somewhat amorphous. Not all of the items are amulets, and not all of them are Jewish. Some are Samaritan, and others may be Christian or Islamic. There is at least one Masonic medallion. Two may be prehistoric spearheads (GA 16 and 57). It is necessary to distinguish between the primary and secondary uses of some items. There are three wimpels (GA 14), for example; the primary function of these was to bind up a Torah scroll to keep it closed, but they could be taken off the scroll and wrapped around an infant for protection, and thus become amulets. Nor is it impossible that the spearheads, though authenticated as genuinely ancient, were used secondarily as amulets and functioned as a magical knife, known as a Kreissmesser, or – in Alsace – as a Krassmesser, which was used to protect parturient women against Lilith. We took a maximal approach to the collection, and digitised and catalogued everything that has a Gaster Amulet number. The basic records were prepared by Gal Sofer of the Ben Gurion University of the Negev, Beersheva, Israel, apart from a small collection in Arabic script, which was prepared by Dora Zsom of the Eötvös Loránd University, Budapest. These records were reviewed and where necessary supplemented by Zsófia Buda, all under the direction of Philip Alexander, the principal investigator (PI) for the project.
2. There is as yet no adequate biography of Gaster. Derek Taylor, *Haham Moses Gaster: Wayward Genius* (London: Vallentine Mitchell, 2021) provides the basic story of his life, but is woefully inadequate on his scholarship. The best overview of his collecting is Maria Haralambakis [Cioata], 'A Survey of the Gaster Collection at the John Rylands Library, Manchester,' *Bulletin of the John Rylands Library*, 89:2 (2013), 107–30.
3. Manchester Digital Collections: www.digitalcollections.manchester.ac.uk/collections/hebrew/1; KTIV: www.nli.org.il/en/discover/manuscripts/hebrew-manuscripts [accessed 3 June 2025]
4. Genizah collection in Library Digital Collections: https://luna.manchester.ac.uk/luna/servlet/ManchesterDev~95~2 [accessed 3 June 2025]; Friedberg Genizah Project: https://fgp.genizah.org/ [accessed 3 June 2025]. To see the Rylands Genizah collection, choose 'Manchester' as the library. See also Renate Smithuis and Philip Alexander (eds), *From Cairo to Manchester: Studies in the Rylands Genizah Fragments* (Oxford: Oxford University Press, 2013). Renate Smithuis played a crucial and substantial role in the cataloguing of the manuscripts and the Genizah fragments, both as researcher and, for a time, as PI.

5   Gaster Amulet Collection in MDC: www.digitalcollections.manchester.ac.uk/collections/gasteramulets/1 [accessed 3 June 2025].
6   For the blogs, see https://rylandscollections.com/ [accessed 3 June 2025]. For the online exhibition, see https://www.digitalexhibitions.manchester.ac.uk/s/jewish-manuscripts/page/introduction [accessed 3 June 2025].
7   See https://collections.vam.ac.uk [accessed 24 January 2025].
8   The Magen David is now inextricably linked with Judaism due to its use on the Israeli flag, but its adoption by Jews is actually post-medieval. For Jewish amulets in general, see Joshua Trachtenberg, *Jewish Magic and Superstition: A Study of Folk Religion* (1939; reprinted Philadelphia: University of Pennsylvania Press, 2004); Gideon Bohak, *Ancient Jewish Magic: A History* (Cambridge: Cambridge University Press, 2008). More specifically, see T. Schrire, *Hebrew Amulets: Their Decipherment and Interpretation* (London: Routledge & Kegan Paul, 1966), and Eli Davis and David A. Frenkel, *The Hebrew Amulet: Biblical-Medical-General* (Jerusalem: Institute for Jewish Studies, 1950) [in Hebrew]. Two exhibition catalogues contain a rich selection of images: Filip Vukosavović (ed.), *Angels and Demons: Jewish Magic through the Ages* (Jerusalem: Bible Lands Museum Jerusalem: Jerusalem, 2010), and Gideon Bohak and Anne Hélène Hoog (eds), *Magie: anges et démons dans la tradition juive* (Paris: Musée d'art et d'histoire du Judaïsme, 2015). The collection featured in the Paris exhibition is in many ways similar to that in the Rylands; one reason for this is that both collections are heavily North African in provenance. See also the amulets in the Gross Family Collection, hosted online on the website of The Centre for Jewish Art at the Hebrew University of Jerusalem (https://cja.ac.il/gross/browser.php?mode=main) [accessed 3 June 2025].
9   A good example of this is found in Bohak and Hoog, *Magie: anges et démons*, p. 43 (catalogue 29). This shows a necklace (Kurdistan, c. 1920) with four tubular amulet cases and other amuletic items on a single chain. The chain and the amulet cases are of silver. It should be noted that the amulet cases are securely held in position on the chain by hoops. See also the amulet (Baghdad, 1892–3) in the Gross Family Collection at https://cja.huji.ac.il/gross/browser.php?mode=set&id=36976 [accessed 4 February 2025]. It is not easy to see how our containers could have been held in place if they were on the same fabric lanyard.
10 The analysis was carried out on 31 October 2024. We are grateful to Ira Rabin for arranging it and putting her expertise at our disposal.
11 Vukosavović, *Angels and Demons*, pp. 116–17.
12 Bohak and Hoog, *Magie: anges et démons dans la tradition juive*, p. 113.
13 Mezuzot are quite often found in Jewish amulet collections: see, for example, GA 66, the metal container of which is very similar to the metal tubes of GA 34 in both material and construction. See also Bohak and Hoog, *Magie: anges et démons*, p. 64 (catalogue 116) and p. 68 (catalogue 117). Tefillin, however, are not found, perhaps because they are only worn at certain times. Their ancient association with protection is nevertheless attested by their early Greek name *phulaktērion* (phylactery), from the verb *phulassō*, 'to protect'.
14 This technique has been applied to other objects, such as the Herculaneum Papyri and the Ein Gedi Leviticus Scroll. These exist as charred rolls which were so badly damaged

by fire that to attempt to unroll them physically risked them crumbling to dust. X-ray CT played a part in the virtual unrolling. For the Herculaneum Papyri, see Inna Bukeeva et al., 'Investigating Herculaneum Papyri: An Innovative 3D approach for the Virtual Unfolding of the Rolls,' *arXiv.org*, on the Cornell University website (arxiv.org/abs/1706.09883, 2017). For the Ein Gedi Leviticus Scroll, see W. Brent Seales et al., 'From Damage to Discovery via Virtual Unwrapping: Reading the Scroll from En-Gedi,' *ScienceAdvances*, 2:9 (2016). The Petra Papyri also survived only as charred rolls, but they seem to have been physically unrolled. How this was done does not seem to have been published. See J. Frösén, A. Arjava and M. Lehtinen (eds), *The Petra Papyri I* (Amman: American Center of Oriental Research, 2002).

15 Here one may compare what was achieved in the case of the Ein Gedi Leviticus scroll (see note 14).

16 That is to say, apart from the very ancient amulets dug up by archaeologists: see Bohak, *Ancient Jewish Magic*.

17 The Dead Sea Scrolls offer a ready example of high-value cultural artefacts, and this has resulted in an inordinate amount of resources being consumed in their conservation, decipherment and interpretation. Fundamentally, that value is related to the fact that they illuminate Judaism in the time of Jesus, and so contribute to our understanding of the origins of Christianity. Since the original black-and-white photographs of the 1950s they have been subjected to various more advanced photographic techniques (usually to try and counteract their physical deterioration, which has been alarming), but none of them would obviously benefit from X-ray tomography, save perhaps the Copper Scroll. That, however, was successfully opened manually by H. Wright Baker, Professor of Mechanical Engineering at UMIST Manchester, in 1956, using decidedly low-tech means involving an adapted dental drill! See H. Wright Baker, 'Notes on the Opening of the "Bronze" Scrolls from Qumran', *Bulletin of the John Rylands Library*, 39:1 (1956), 45–56. It is interesting to speculate how X-ray tomography would have coped with the problem of the *unrolled* Copper Scroll. It is also interesting to speculate whether such a high-risk, invasive procedure as Wright Baker applied would be countenanced today.

Manchester University Press

The *Apocalypse* and *Biblia pauperum* Blockbooks Bound by Johannes Richenbach in 1467

STEPHEN MOSSMAN, UNIVERSITY OF MANCHESTER
EDWARD POTTEN, UNIVERSITY OF MANCHESTER

Abstract

This study examines a composite volume (Manchester, John Rylands Library, S16119) bound by Johannes Richenbach in 1467, which contains two blockbook editions: a coloured *Apocalypse* (edition IV D) and an uncoloured *Biblia pauperum* (edition III). The binding, executed for the Franciscan friar Ulrich Geislinger, provides a rare, securely dated piece of evidence for the production of these editions. We remove historical scepticism about the authenticity of this binding through analyses of the watermarks present in the paper stocks and the material composition of the printing inks. These analyses are combined to assess the geographical and chronological origins of the blockbooks. The *Biblia pauperum* proves to be a Netherlandish product datable to *c*. 1460–3, while the *Apocalypse* cannot postdate *c*. 1462/3 and is probably rather earlier. The study assesses the value of these analytical techniques to understanding the geographical and chronological contours of the production of xylographic books in the fifteenth century.

Keywords: blockbooks; *Apocalypse*; *Biblia pauperum*; Johannes Richenbach; watermarks; ink analysis

On 7 October 1802, the Benedictine monk, diplomat, spy and book agent Alexander Horn (1762–1820) wrote to one of his greatest patrons, George John, the second Earl Spencer (1758–1834) about what would become one of Spencer's greatest acquisitions. He described:

> ... a very curious article which I bought a few days ago at Ulm consisting of the *Biblia pauperum* and *Historia S. Ioannes Apocalyptici* in their original old binding with the following inscription printed or impressed on the sides of the binding in the following manner:
>
> Iste liber pertinet ad me /: name :/ vicarium in Ulm anno Domini Mccccxlv.[1]

The volume that Horn was describing is a composite volume (or *Sammelband*), now Manchester, John Rylands Library S16119. It consists of two blockbook editions bound together: first a coloured *Apocalypse*, in edition IV following W. L. Schreiber's definitive categorisation, and then an uncoloured *Biblia pauperum*, in edition III. Several facts had become a little garbled in Horn's memory, including the precise details of the lettering, which reads on the front cover 'iste liber est f[rat]ris v̊lrici gyslinger lectoris i[n] vlma mino[rum]', and on the rear 'p[er] me ioha[n]nem richenbach de gyslingen illigatus est an[n]o d[omi]ni m.cccc.lxvii'. The stamped inscription reveals that the binding was executed in 1467 for the Franciscan friar Ulrich Geislinger (d. 1493), lector of the convent in Ulm; a native of Geislingen an der Steige, he would subsequently be consecrated as an auxiliary bishop in 1474 and accorded the titular bishopric of Adramyttium (modern Edremit).[2] The binder was Johannes Richenbach (d. 1486), a prebendary priest of the parish church in Geislingen an der Steige and arguably the most famous German binder of the fifteenth century.[3] The survival of any one of these three objects (i.e. the two blockbooks and their binding) would be remarkable; the survival of all three still in situ is doubly so. Fifty-eight or fifty-nine Richenbach bindings survive, depending on how one counts a two-volume set. This is the only blockbook among them.

The date of 1467 on the binding gives the volume a particular significance. Blockbooks – books printed in relief, with text and image cut into a series of wooden blocks – are very rarely signed or dated by either the woodcutter or the printer. Unlike typographic books, where the type was set and an entire edition printed in one campaign before the formes were dismantled, requiring the type to be reset anew to print a further edition, xylographic books could be printed on demand in small runs, and the blocks returned to the shelf to be taken down again when more copies were required. Impressions might be taken across several decades, complicating the notion of an 'edition', and sets of blocks could be sold and used in a different location. This makes placing blockbooks in space and time difficult. The Richenbach binding provides one of very few pieces of securely dated evidence for the production of these *Apocalypse* and *Biblia pauperum* editions: a *terminus ad quem* of 1467, before which both copies must have been printed. The scarcity of such evidence is well illustrated by *Apocalypse* edition IV. Some thirty-four copies of this edition survive – an unusually large number, and, indeed, about one in twelve of the entire extant corpus of fifteenth-century blockbooks – yet only two bear any dating evidence at all. This binding is the first. It is considered the more secure: the crucial piece of certain information on which all else depends. The second is just a note (in Mainz, Gutenberg Museum, Ink. 131, fol. 48v) that in 1463 its author had entered the service of Heinrich III, Landgraf von Oberhessen-Marburg (r. 1458–83); since this could have been added at any point after 1463, it need have no particular bearing on the dating of the blockbook itself.[4]

Long considered an essential intermediary stage between single-sheet xylographic (woodblock) printing in Western Europe and the invention of printing with moveable type, blockbooks assumed a venerated position among book collectors and historians from the late sixteenth century onwards. They were weaponised in debates about the precise location of the invention of printing, which were dominated by arguments for or against Laurens Jansz. Coster in Haarlem. Blockbooks were central to these arguments, understood as a 'missing link' in the story of the origins of European printing. Consequently, they were eagerly sought out by collectors, most of whom had them removed from bindings that they viewed as old and uninteresting, the leaves washed to remove traces of earlier ownership and the books rebound in brightly coloured crushed Morocco. This history of 'trophy collection' leaves scholars interested in the codicology of blockbooks with a serious evidentiary problem, which Nigel F. Palmer succinctly summarised: 'the process of washing, pressing, repair, remounting of the leaves, and rebinding to which these prized items were subjected [...] has destroyed 90 percent of the evidence for the original assembly and codicological presentation of medieval blockbooks. It is hard to think of another category of book for which the codicological evidence has been so comprehensively destroyed.'[5]

Although it is easy to be critical of the bibliophiles, it is worth noting that the survival rate for blockbooks in early bindings may have been relatively low even in the late eighteenth century, when so many were being brought to light. Andrew Honey's work on the Bodleian blockbooks indicates that due to the very nature of their production – printed anopisthographically (i.e. on one side of the paper only) with very narrow inner margins, then assembled either as stacked sheets, with the individual leaves folded and arranged as a series of individual bifolia, or alternatively with the leaves folded and gathered into three or four quires – they may have been better suited to a tacketed style of binding, where the long stitching allowed them to open well without putting strain on the joints. Those which were not bound in this style would potentially have succumbed to damage along the stitched folds.[6] Furthermore, we know very little about how blockbooks were used in the fifteenth century, but it is certain that some at least were not immediately bound as codices. A *Biblia pauperum* in Oxford (Bodleian Library, Douce 248) has nail holes at each corner of the leaves, and shows no trace that it was ever sewn, suggesting that in this instance the sheets were pinned to a wooden board – perhaps on a wall or pew – before later being dismounted and affixed to guards in order to be bound as a book.[7] All of these factors coincide to make blockbooks in contemporary bindings extremely rare. The authors are aware of just eleven examples in early bindings, two of which are the focus of this article.

The scarcity of blockbooks surviving in their fifteenth-century context has always led to a degree of scepticism about the Rylands Richenbach binding, which seems almost too good to be true. This scepticism began with a spat between Horn and Thomas Frognall Dibdin (1776–1847). Writing in 1814 in his monumental catalogue of Earl Spencer's library, Dibdin accused Horn of breaking up the Richenbach *Sammelband* and selling the contents separately.[8] Horn was incensed, and wrote to Spencer on 22 February 1818:

> I beg now leave to state an inaccuracy of Mr. Dibdin's in your Lordship's first vol. which reflects on myself, and which I should not have noticed if it had not been severely censured in the German Journals. I had told that Gentleman that I had formerly been in possession of 3 [sic] Block books bound up together in their original binding [...] This book [I] informed Mr Dibdin I had sold to Mr. Edwards for a large sum pointing out the high importance of the old binding, and requesting him not to separate them; but after all I am accused by Mr. D of having committed that vandalic crime.[9]

Horn was sufficiently outraged to make note of the slander in the pages of the *Allgemeine Literatur-Zeitung*.[10] Although Horn had first offered the volume to Spencer in 1802, Spencer's offer was not sufficient to secure it, and Horn instead sold it to the bookseller James Edwards (1757–1816). Edwards in turn sold it to the Neapolitan bibliophile and collector Luigi Serra di Cassano, fourth duke of Cassano (1747–1825).[11] Cassano sold his entire library to Earl Spencer in 1819/20, and so, nearly twenty years after the initial offer, the blockbooks bound by Richenbach finally ended up at Althorp.

Suspicion about the volume remained, however, and it was sufficiently widespread as to require refutation in Samuel Leigh Sotheby's 1858 *Principia typographia*:

> Since we examined this volume some few years ago, the two works have been taken out of the old covers and re-inserted, in consequence of many of the leaves being loose. As this circumstance might warrant any person hereafter to assert, that the copies had been placed in the old binding as a matter of deception, we mentioned our fears to Mr. Appleyard, the librarian of Earl Spencer. Mr. Appleyard at once saw the force of our argument, and immediately placed in our hands the annexed account from the binder.

> "The Right Hon. Earl Spencer.   London, Feb. 5th, 1850.
> To CLARKE & BEDFORD, Bookbinders, 61, Frith Street, Soho Square.
> Historia St. Joannis & Biblia Pauperum, 4to. The whole taken out of Old Covers, the Biblia Pauperum cleaned, and worm holes mended, both works, mounted on guards and rebound into old Covers again, and mending in Ditto. £2 12 0."[12]

Despite this public announcement, suspicion remained, perhaps because of the slightly odd way in which Clarke & Bedford undertook the rebinding. As Honey has noted, blockbooks are not easy to bind, and the binder in 1850 mounted the individual leaves on to guards, then replaced the original sewing entirely. This has resulted in a binding which looks misshapen; the hinges and spine 'feel' wrong. It gives the impression of a book that once held a text block considerably deeper than that which it now surrounds.

In the AHRC/DFG-funded project '*Werck der bücher*: Transitions, Experimentation, and Collaboration in Reprographic Technologies, 1440–1470', we aim to assemble the totality of evidence that can be provided by datable watermarks in the paper stocks of all surviving blockbooks. Our ultimate goals involve the assessment of that evidence seen in its entirety: to chart the chronology of xylographic print production and to understand its intersection, grounded in the evidence of shared paper stocks, with typographic printing in the same period. Findings significant for each copy are generated in the process. The blockbooks in Manchester, which have never been subjected to sustained scholarly scrutiny, are to be studied with especial intensity. Here we undertake analysis of the watermarks in the paper stocks, which we can examine with particular precision using high-resolution imagery, and of the inks used to print the two blockbooks that Richenbach bound in 1467. This allows for their precise dating; since neither can have been produced later than c. 1463, any residual doubt as to whether these really were the books that Richenbach sewed into this binding is allayed. The information that the *Biblia pauperum* (but not the *Apocalypse*) was cleaned in 1850, meanwhile, is essential to understand the results of the ink analysis.

We begin with watermark evidence, and with the *Biblia pauperum* edition III (16119-2). Printed anopisthographically, uncoloured and with the blank sides not pasted together, particularly sharp images of the watermarks have been obtained using transmitted and reflected light images, recombined digitally using a method developed at the John Rylands Library by Tony Richards.[13] It is difficult to know whether the blank sides were once pasted or bore other traces of use, because the book was so intrusively washed in 1850. Certainly, however, since this edition of forty leaves was printed from twenty blocks so as to be made up into two quires of ten bifolia, it was conceived from the outset as a booklet (only those blockbooks printed as 'stacked sheets' can have had an alternative – perhaps primary – deployment as a poster cycle). Editions I, II and III belong to the first group of Latin *Biblia pauperum* editions. Edition III is distinguished by its use of sixteen new blocks, re-cut on the model of edition I, together with four blocks that had already been introduced in edition II, where they accompanied sixteen of the original blocks from edition I.[14] Whether this 'first' group is actually the earliest in chronological sequence is uncertain. The ultimate point of conception in both chronological and geographical terms for the forty-leaf Latin *Biblia pauperum* depends on the exact interpretation of close parallels identified with Utrecht manuscript illumination dated 1460.[15]

There are four different watermarks present in the paper stocks of 16119-2; all are anchors, atop which is mounted a single-contoured cross. The first and second, present in a single example each (leaves 2 and 7; Figures 1 and 2), form a pair.[16] Watermarks come in pairs within a single ream of paper, as is well known, because two moulds were used concurrently at a single vat of pulp.[17] This pair is of a type that is only poorly documented in our reference corpora. It is most similar to Piccard, *Anker*, II 711 (attested in Kleve, 1463), but little weight can be placed on that date given the degree of dissimilarity. The third watermark (leaves 4 (Figure 3), 5, 9, 12, 13, 28, 30, 35, 37 and 40), which may very well form a pair with the fourth – but which we treat separately here for the purposes of identification – gives us greater security.[18] It belongs to a taxonomic group defined by Piccard, *Anker*, II 499–506 (attested principally, moving from west to east, in Utrecht, Culemborg, Arnhem and Kleve in 1454–63), and is most similar to II 506 (attested in Kleve, 1463). The fourth watermark (leaves 16 (Figure 4), 18, 19, 22, 24, 25, 31 and 34) belongs to the same taxonomic group and is most similar – although with many more 'dots' – to II 505 (attested in Utrecht, 1461).[19] We have not found additional comparators in the online databases that augment Piccard's published examples of these types.[20] Here, then, there exists strong cumulative evidence for production of this copy in or around the Dutch towns along the rivers Rhine and Lek in c. 1460–3. The degree of caution that we should exercise with regard to these chronological parameters is difficult to estimate. Under normal circumstances, paper of standard sizes was routinely used within two to three years

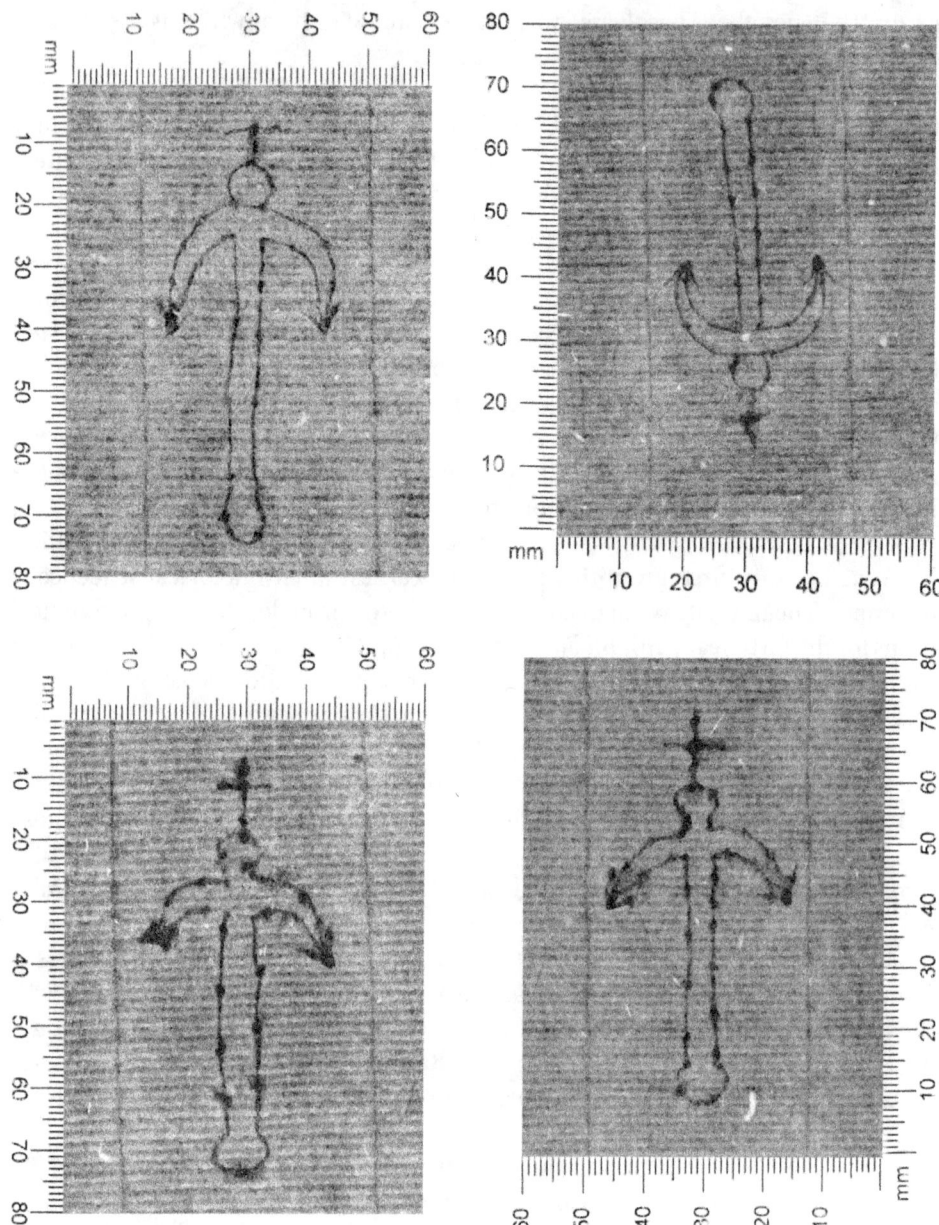

Figures 1-4 Manchester, John Rylands Library, 16119-2, watermarks in, clockwise, fols 2, 7, 16 and 4. © The John Rylands Research Institute and Library, University of Manchester.

of production.[21] That might lead one to greater conservatism in relating known dates of use to undated items (especially in the absence of exact matches). Yet it must also be kept in mind that paper for printing had to be used 'fresh': cases are documented of fifteenth-century typographic printers rejecting shipments of paper that had lain in storage for too long, thus having become too brittle and unable to be sufficiently dampened again to accept an inked impression well.[22]

The *Apocalypse* edition IV presents a different set of problems. It was printed on forty-eight leaves from twenty-four blocks so as to be made up into three quires of eight bifolia: like the *Biblia pauperum* edition III, it was conceived from the outset as a booklet. Edition IV can be subdivided into five states, labelled A–E and distinguished by the cumulative presence of alterations made to the blocks. The Manchester copy (16119-1) is one of just two surviving copies of edition IV D (the other is New York, Pierpont Morgan Library, PML 1051).[23] The leaves, printed anopisthographically, have been pasted together in this copy on their blank sides, and such heavy colour has been applied that the combination of these two circumstances has made it impossible to obtain good images with transmitted and reflected light. Only by means of the older method of beta-radiography, or ideally with a thermographic camera, would it be possible to obtain images.[24] In this case, one must resort to the oldest of all methods: examination by eye with a strong light aimed directly at the watermark, and a ruler.

The paper stock is watermarked, as Allan Stevenson first noted, with the arms of Metz (a shield party per pale, sable and argent) enclosed within a circle.[25] Reliant on Stevenson's statement, Paul Needham claimed in 2013 that the same paper stock had been used to print a copy (New York, Pierpont Morgan Library, PML 6) of an earlier state of the same edition (*Apocalypse* IV B).[26] If these two books, representatives of two different and non-consecutive states of an edition, really were to share an identical paper stock, then an intractable problem of dating would emerge. In fact, comparison of the watermarks in 16119-1 with beta-radiographs of those in PML 6 reveals that this is not so.[27] The pair of watermarks in PML 6 (state IV B) uses a much denser grid of wire mesh to indicate the sinister sable than does the corresponding pair in 16119-2 (state IV D). This is paper from the same mill, but these are not the same watermarks; they are two different pairs, produced by different sets of moulds. The type is not one that is well documented in the reference corpora. Charles-Moïse Briquet published a single example, of which he had found variants in documents dated 1448–61.[28] Nikolai Likhachev published a further four: three from 'Slavonic manuscripts' of the mid-fifteenth century, and one in 'a Russian document' dated *c.* 1450–62.[29] Piccard recorded eleven examples located in documents dated 1451–60.[30] All of these examples, with the exception of the Russian material, were found in locations along an arc from the southern Low Countries through the lower Rhineland into Lorraine. This is paper, as Briquet noted, from the civic mill in Metz; and of that mill, two recent studies have considerably extended our knowledge well beyond Briquet's introductory remarks.[31]

The paper mill of Metz was founded by the city government in 1446/7. It was operated as a state-run enterprise until it ceased production in 1459, after the second of two successive contractors had failed to turn it into a going concern; it was ultimately privatised in 1461/2. For this period of nearly fifteen years, its accounts were run through the central state accounts for the city, and in consequence they survive. This allows the operation of the mill – including the precise grades of paper produced and their cost – to be charted in minute detail. The paper used for blockbooks was presumably that of the best quality (*boin pauppier*), intended for writing (as opposed to wrapping and packaging), and sold consistently at 9 s. per ream of 500 leaves. Although production ceased in 1459, probably in March of that year, the sale of residual unsold stock was not complete until the financial year 1461/2.[32] The accounts record some seventy-one transactions of sale to more than fifty named purchasers, some of whom were regular customers. Most were small-scale purchases made locally (29 per cent were of fewer than five reams, and 53 per cent of five to eight reams), and the city administration used paper from the mill as of 1448, but the largest quantities went to merchants selling Metz paper farther afield. Over half the total volume sold is accounted for by just 18 per cent of the total transactions, including one very large sale of 750 reams.[33] Onward transportation was organised at civic expense for large shipments of paper to Antwerp in 1451/2 and 1461/2, the latter of *c.* 300–480 reams, thence for resale in the Low Countries.[34] The moulds (and thus their watermarks) must have been made by the two consecutive master paper-makers in the mill themselves, since there is no record in the civic accounts – which otherwise detail expenditure on raw materials and equipment precisely – for the purchase of moulds (the trade of the specialist mould-maker seems to have been a later development).[35]

Neither of these two studies, however, has done more than to report Briquet's information on the watermarks from the Metz mill.[36] There is, at least, no documentary evidence that the four or five different grades of paper that it produced were distinguished, as was otherwise usual, by the use of different watermarks.[37] Without taking the step that Stevenson proposed – to scrutinise the full run of civic archival documentation in Metz for the relevant period in order to locate exact dated parallels – neither of the *Apocalypse* edition IV blockbooks printed on paper watermarked by the arms of Metz can be more precisely dated.[38] The examples noted by Briquet, Likhachev and (more instructively) Piccard do not permit the identification of exact matches. One might note, however, that the denser grid of wire mesh used for the sinister sable in the watermarks of PML 6 can be found on just three of Piccard's eleven examples, all dated 1454–5 (PO-24768, 24772 and 24778), and these may well correspond to Briquet's 'autre var., un peu plus petite, avec les lignes hàchées plus serrées', for which he found dated examples 1455–9. If we accept a dating to the mid- or late 1450s for this copy of the *Apocalypse* edition IV B, then the Manchester copy of edition IV D must necessarily be more recent. It is very unlikely to postdate *c.* 1461/2; it is worthy of note that although unsold stock produced in or before 1459 was still being sold off at that point, neither Briquet nor Piccard found any dated instance of this paper in use more recently than 1461.[39] We must bear in mind again that paper which had been stored for a long period seems to have been found unsuitable as a substrate for printing.

Ad Stijnman's seminal (and hitherto only) study of the inks used to print blockbooks first deployed a combination of X-ray fluorescence (XRF) spectrometry and near-infrared (NIR) reflectography to examine a single blockbook leaf, and then, in a second step, matched up the observations recorded with the information provided in late medieval ink recipes to identify the material composition. Stijnman argued that blockbook inks could be divided into three main groups: carbon-based inks made from lampblack (i.e. those also used by typographic printers), which remain consistently black over time; iron-gall inks, with gum Arabic added to increase viscosity and improve adherence, which become browned over time; and iron-gall inks with lampblack added, which become browned over time, but much more slowly.[40] To examine the inks used to print the two present blockbooks, we first followed the protocol established by Ira Rabin for the study of historic inks by NIR and ultraviolet (UV) reflectography at two specific wavelengths, using a Dino-Lite AD4113T-12V digital microscope.[41] Rabin's observations regarding the appearance of iron-gall manuscript inks under microscopic examination, which 'appear highly inhomogeneous in colour and texture, with traces of dark crystals', and the wavelength at which they become fully invisible under infrared light (only above $c.$ 1400 nm), may call into question Stijnman's interpretation of the dark crystalline particles observed in the iron-gall ink used to print the blockbook leaf that he examined as carbon-based lampblack.[42] A resolution to that issue, however, must await further study.

The *Apocalypse* edition IV D blockbook (16119-1), when inspected microscopically and by UV reflectography at 395 nm, showed that same inhomogeneous particulate composition. Under NIR reflectography at 940nm, the browned ink appeared as a light grey, but remained visible (Figure 5). This replicated Stijnman's findings exactly. This is no surprise, since the single leaf that he examined (Amsterdam, Rijksmuseum, RP-P-2009-24) is also a fragment of an *Apocalypse* edition IV. The *Biblia pauperum* edition III blockbook (16119-2), however, and contrary to our expectation, became entirely invisible when subjected to NIR reflectography at 940 nm (Figure 6). This behaviour is less what one would expect of an iron-gall ink and more akin to that of a tannin (plant-based) ink, which disappears at $c.$ 750nm. Yet subsequent XRF reflectography undertaken by Ira Rabin, using a Bruker ELIO spectrometer, revealed that the material composition of the two inks was similar. Both are, in fact, iron gall inks. The *Apocalypse* blockbook ink showed a strong iron signal, with copper and lead also present (given that their intensity was higher than that measured in the surrounding paper), although this may have arisen from contamination (Figure 7). The *Biblia pauperum* blockbook ink showed an equally strong iron signal (Figure 8). The change in visibility observed during NIR reflectography is considered by Ira Rabin (personal communication), given the same formulation of the ink, to be a consequence of a different degree of deterioration of the ink. That in turn is very probably resultant from the washing of the *Biblia pauperum* blockbook, especially if an acidic reagent was used in 1850. A striking feature of the *Biblia pauperum* analysis was the very strong lead signature in both paper and ink in one part of the leaf but not in another: this is, at present, inexplicable and its cause unknown.

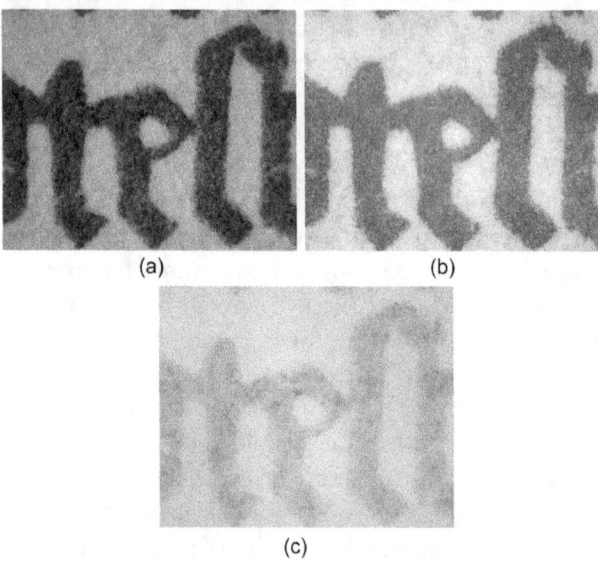

Figure 5 Manchester, John Rylands Library, 16119-1, fol. 3r, using Dino-Lite AD4113T-12V microscope at magnification 43.8: (a) under visible light, (b) UV reflectography at 395nm and (c) NIR reflectography at 940 nm. © Stefan Hanß and Stephen Mossman.

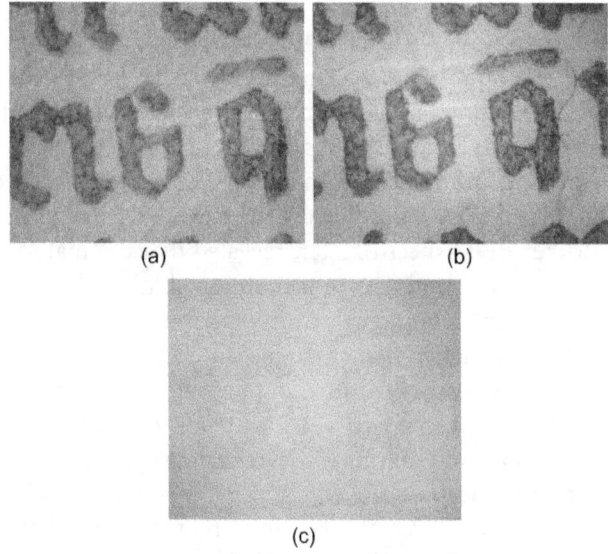

Figure 6 Manchester, John Rylands Library, 16119-2, fol. 2r, using Dino-Lite AD4113T-12V microscope at magnification 43.8: (a) under visible light, (b) UV reflectography at 395nm and (c) NIR reflectography at 940 nm. © Stefan Hanß and Stephen Mossman.

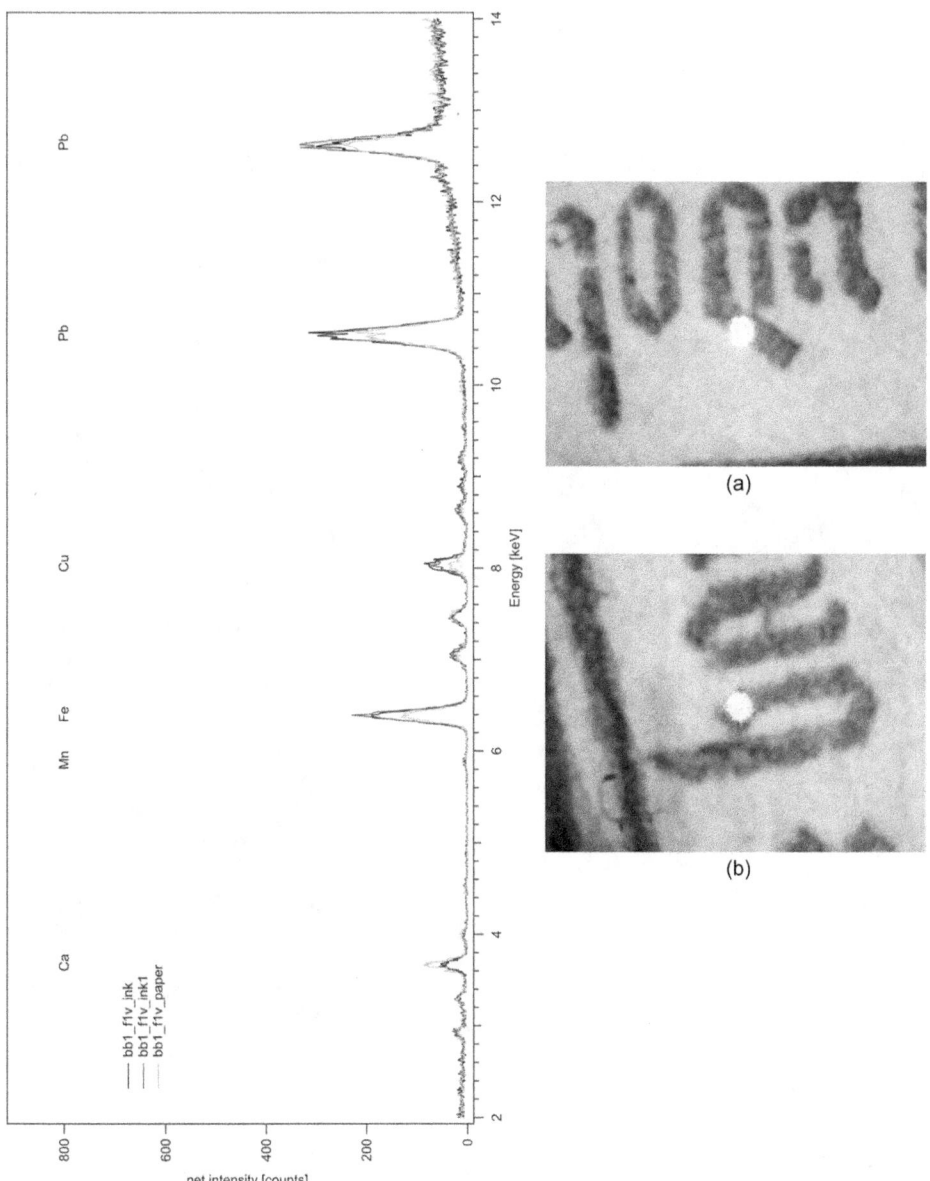

Figure 7 XRF spectrum and spot identifications (from fol. 1v) for Manchester, John Rylands Library, 16119-1 (*Apocalypse* edition IV D). XRF analysis by Ira Rabin (BAM) using Bruker ELIO spectrometer with parameters 40kV, 80 μA, 120 s.

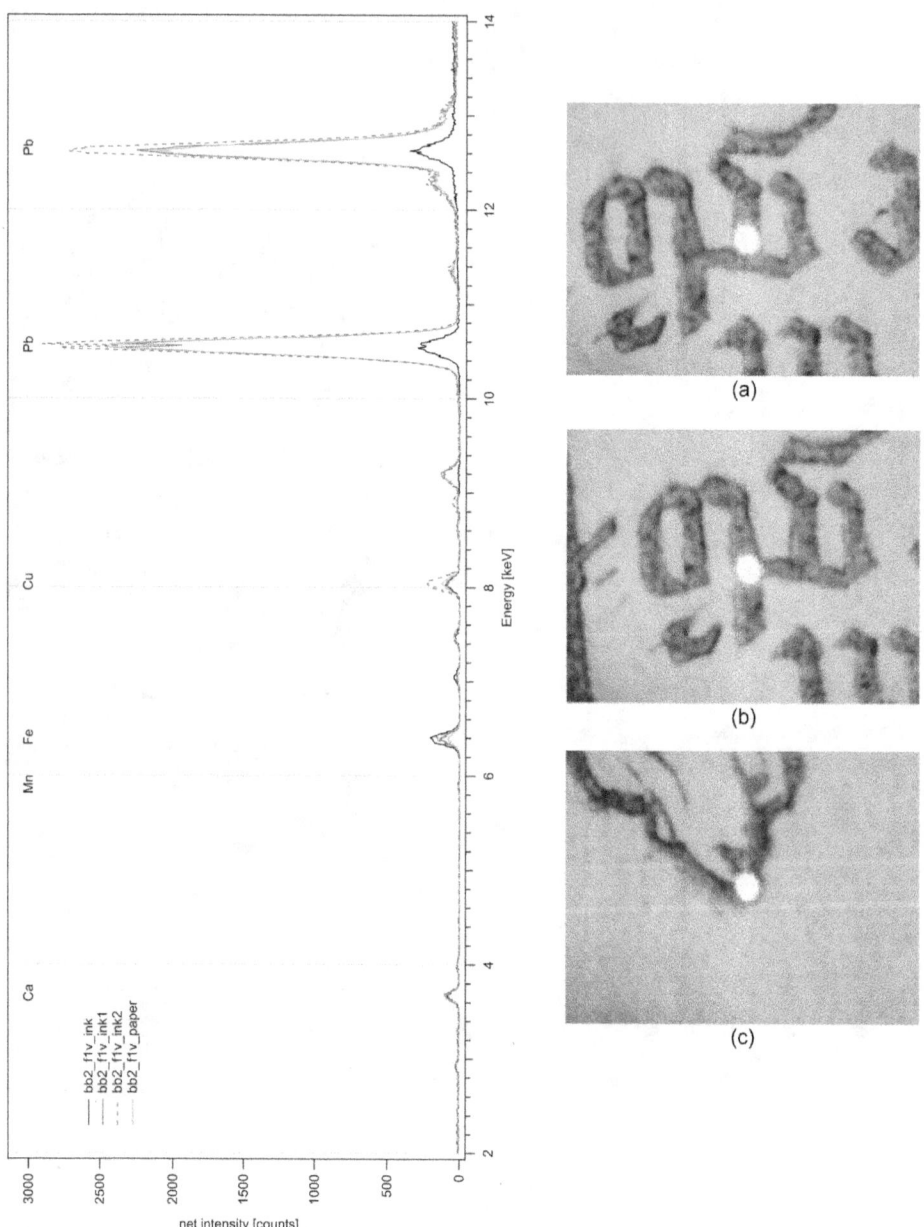

Figure 8 XRF spectrum and spot identifications (from fol. 1v) for Manchester, John Rylands Library, 16119-2 (*Biblia pauperum* edition III). XRF analysis by Ira Rabin (BAM) using Bruker ELIO spectrometer with parameters 40kV, 80 μA, 120 s.

The material analysis of these two blockbooks gives us no reason to think that these were not the books to which Johannes Richenbach applied the present binding in 1467. Despite the curious appearance of its spine, and the doubt that has been variously cast upon the authenticity of the book, there are no evidentiary grounds to suspect that it is a *remboîtage*. Whether this was the first time that these books were bound is a different matter. It is conceivable that they were stitched initially into a limp binding, which would have suited their format better, since – contrary to repeated assumption in the scholarly literature – they were not printed shortly before 1467, but at least four years beforehand; nor were they locally produced. The *Biblia pauperum* (16119-2) is a Netherlandish product, datable with reasonable security to *c.* 1460–63. The *Apocalypse* (16119-1) is less securely datable, but does not postdate *c.* 1462/3 and may well be somewhat earlier. The geographical location of *Apocalypse* edition IV has been the subject of some debate. It is considered to be German, but the pertinent evidence sheds light almost exclusively upon reception, not production, and does not concentrate in a closely defined regional pattern.[43] The known geographical reach of paper from the short-lived state-owned mill at Metz provides an important indication of its production – if we hold the edition to be German and rule out the southern Low Countries – in the lower or central Rhineland. This would concur with Palmer's assessment of the paper stocks used in a late group of copies of *Apocalypse* edition IV E, 'which is to be placed somewhere in western Germany with access to paper stocks from the Vosges, possibly in the region of Cologne, shortly before or even after 1470'.[44] The ink analysis is inconclusive in its ultimate significance for these questions, since the material composition of the two inks used proves to be too similar to conclude that they must definitely represent the ink recipes of two different workshops.

Acknowledgements

The authors are extremely grateful to Tony Richards (John Rylands Research Institute and Library) for his imaging of the watermarks, and to Stefan Hanß (University of Manchester) and Ira Rabin (Bundesanstalt für Materialforschung und -prüfung, Berlin) for their analysis of the printing inks, to provide the evidentiary basis for this present study.

Notes

1 London, British Library, Althorp papers, Add. MS 75965. The authors are very grateful to Eric White for sharing his research notes on Horn's discovery of the Richenbach binding and the subsequent dispute between Horn and T. F. Dibdin. White notes (personal communication) that many of Horn's acquisitions in 1802 came from the Charterhouse at Buxheim, perhaps indicating the source of this present *Sammelband*.

2 Klaus Graf, 'Kunden des Geislinger Buchbinders Johannes Richenbach im 15. Jahrhundert: Ulrich Geislinger, Weihbischof von Augsburg', *Archivalia*, 21 April 2019, https://doi.org/10.58079/c953.

3 The authoritative source for Richenbach and his bindings is, in fact, https://de.wikipedia.org/wiki/Johannes_Richenbach [accessed 14 October 2024], on which see Klaus Graf, 'Johannes Richenbach, der vielleicht berühmteste deutsche Buchbinder des 15. Jahrhunderts', *Archivalia*, 27 May 2019, https://doi.org/10.58079/c9at.

4 Rahel Bacher, 'AP-04: Johannes ‹Apostolus› Apokalypse ‹lat.› (Apocalysis). Ausgabe IV', in Bettina Wagner (ed.), *Xylographa Bavarica. Blockbücher in bayerischen Sammlungen (Xylo-Bav)*, Bayerische Staatsbibliothek. Schriftenreihe 6 (Wiesbaden: Harrassowitz, 2016), pp. 118–20; Elke Purpus, *Die Blockbücher der Apokalypse*, Edition Wissenschaft. Reihe Kunstgeschichte 20 (Marburg: Tectum, 1999), pp. 164–6; *Die Apokalypse: Blockbuch-Ausgabe IV E. Mainz, Gutenberg-Museum, Ink. 131. Farbmikrofiche-Edition*, ed. Elke Purpus, Monumenta xylographica et typographica 1 (Munich: Edition Helga Lengenfelder, 1991), pp. 22–3.

5 Nigel F. Palmer, 'Woodcuts for Reading: the Codicology of Fifteenth-Century Blockbooks and Woodcut Cycles', in Peter Parshall (ed.), *The Woodcut in Fifteenth-Century Europe*, Studies in the History of Art 75. Center for Advanced Study in the Visual Arts Symposium Papers 52 (New Haven, CT and London: Yale University Press, 2009), p. 94.

6 Andrew Honey, '»The binding was the ancient legitimate one«: Looking for Early Binding Evidence in Blockbooks', in Bettina Wagner (ed.), *Blockbücher des 15. Jahrhunderts: Eine Experimentierphase im frühen Buchdruck*, Bibliothek und Wissenschaft, 46 (2013), 83–5.

7 Nigel F. Palmer, with binding descriptions by Andrew Honey, 'Blockbooks, Woodcut and Metalcut Single Sheets', in Alan Coates et al., *A Catalogue of Books Printed in the Fifteenth Century Now in the Bodleian Library*, vol. 1 (Oxford: Oxford University Press, 2005), pp. 14–16 (BB-4); Palmer, 'Woodcuts for Reading', pp. 101–4.

8 Thomas Frognall Dibdin, *Bibliotheca Spenceriana*, vol. 1 (London: Shakspeare Press, 1814), pp. iv–v n. †.

9 London, British Library, Add. MS 75965.

10 *Allgemeine Literatur-Zeitung* (March 1818), no. 56, p. 448.

11 [Gabriele Stasi], *Catalogo dell'edizioni del sec. XV. esistenti nella Biblioteca del Duca di Cassano Serra* (Naples: n.p., 1807), p. 8: 'Biblia Pauperum, adjic. in calce Historia S. Joannis, formis ligneis expressa fol. Ambedue queste opere, primi sforzi dell'arte d'incidere, sono di una estrema rarità'.

12 Samuel Leigh Sotheby, *Principia typographia* (London: W. McDowall, 1858), vol. 1, p. 22 n. *.

13 Tony Richards, 'Imaging Watermarks', *Rylands Blog*, 14 December 2023, https://rylandscollections.com/2023/12/14/imaging-watermarks/ [accessed 31 January 2025].

14 Heike Riedel-Bierschwale, 'BP-03: Biblia pauperum ‹lat.› (Armenbibel)', in Bettina Wagner (ed.), *Xylographa Bavarica. Blockbücher in bayerischen Sammlungen (Xylo-Bav)*, Bayerische Staatsbibliothek. Schriftenreihe 6 (Wiesbaden: Harrassowitz, 2016),

pp. 143–5; Renate Kroll, 'Beobachtungen zur Ausgabenfolge der 40blättrigen Biblia pauperum', in *Blockbücher des Mittelalters: Bilderfolgen als Lektüre. Gutenberg-Museum, Mainz, 22. Juni 1991 bis 1. September 1991* (Mainz: Philipp von Zabern, 1991), pp. 289–310.

15 Nigel F. Palmer, ed., *Apokalypse – Ars moriendi – Biblia pauperum – Antichrist – Fabel vom kranken Löwen – Kalendarium und Planetenbücher – Historia David: Die lateinisch-deutschen Blockbücher des Berlin-Breslauer Sammelbandes. Staatliche Museen zu Berlin – Preußischer Kulturbesitz, Kupferstichkabinett, Cim. 1, 2, 5, 7, 9, 10, 12. Farbmikrofiche-Edition*, Monumenta xylographica et typographica 2 (Munich: Edition Helga Lengenfelder, 1992), pp. 45–6.

16 Measurements (in mm): (a) leaf 2: width 32.5, height 71.5 (though the top of the cross has broken off), between chain lines of width 40.5; (b) leaf 7: width 36, height 75, between chain lines of width 42. Note the sharp (not rounded) angle where the arms join the shank of the anchor; the narrowness of the arms and their comparative depth; and the way in which the ring at the anchor's crown forms almost a complete circle, not joined to the crown by rounded 'shoulders'.

17 Allan H. Stevenson, 'Watermarks are Twins', *Studies in Bibliography* 4 (1951/2), 57–91 and 235; for the observation that watermarks can, in fact, appear as triplets and quadruplets, not just as twins, with a possible explanation of this phenomenon, see Rahel Bacher and Veronika Hausler, 'Ausgabenübergreifende Auswertung der Wasserzeichen', in Bettina Wagner (ed.), *Xylographa Bavarica. Blockbücher in bayerischen Sammlungen (Xylo-Bav)*, Bayerische Staatsbibliothek. Schriftenreihe 6 (Wiesbaden: Harrassowitz, 2016), p. 27.

18 Measurements (in mm), taken from leaf 4: width 33, height 67, between chain lines of width 42. Note that ours is a somewhat deformed example of its type, with many 'dots', and the flukes on one of the anchor's arms at an odd angle, pointing away from the anchor rather than in line with the shank.

19 Measurements (in mm), taken from leaf 16: width 35, height 67, between chain lines of width 40.

20 Gerhard Piccard, *Wasserzeichen Anker*, Veröffentlichungen der staatlichen Archivverwaltung Baden-Württemberg. Sonderreihe: Die Wasserzeichenkartei Piccard im Hauptstaatsarchiv Stuttgart Findbuch 6 (Stuttgart: Kohlhammer, 1978).

21 Gerhard Piccard, 'Die Wasserzeichenforschung als historische Hilfswissenschaft', *Archivalische Zeitschrift*, 51 (1956), 111–2. When tested, this rule has been found to hold good for the most part, although not exclusively so, for manuscript production: see Alois Haidinger, 'Datieren mittelalterlicher Handschriften mittels ihrer Wasserzeichen', *Österreichische Akademie der Wissenschaften: Anzeiger der phil.-hist. Klasse*, 139 (2004), 14–21. It has held good much more securely for typographic editions, as we might expect from the more rapid use of paper stocks in a print shop: see Theo Gerardy, 'Zur Methodik des Datierens von Frühdrucken mit Hilfe des Papiers', in Hans Limburg, Hartwig Lohse and Wolfgang Schmitz (eds), *Ars impressoria: Entstehung und Entwicklung des Buchdrucks. Eine internationale Festgabe für Severin Corsten zum 65. Geburtstag* (Munich, etc.: K. G. Saur, 1986), pp. 55–60.

22 Maria Zaar-Görgens, *Champagne – Bar – Lothringen: Papierproduktion und Papierabsatz vom 14. bis zum Ende des 16. Jahrhunderts*, Beiträge zur Landes- und Kulturgeschichte 3 (Trier: Porta Alba, 2004), pp. 165–7.
23 Bacher, 'AP-04', p. 119 (with erroneous omission of PML 1051 as a copy of state D); Purpus, *Die Blockbücher der Apokalypse*, pp. 140–7; *Die Apokalypse: Blockbuch-Ausgabe IV E*, ed. Purpus, pp. 18–22.
24 The thermographic method has been developed and refined principally by Peter Meinlschmidt at the Fraunhofer-Institut für Holzforschung in Braunschweig; with particular reference to blockbooks, see Peter Meinlschmidt, Carmen Kämmerer, Volker Märgner and Bettina Wagner, 'Der Einsatz von Infrarot-Technik zur Dokumentation von Wasserzeichen aus Blockbüchern', in Bettina Wagner (ed.), *Blockbücher des 15. Jahrhunderts: Eine Experimentierphase im frühen Buchdruck*, Bibliothek und Wissenschaft, 46 (2013), 13–34. We are not aware, at the present time of writing, of any research library or other institution in the United Kingdom that has acquired a thermographic camera. They represent the gold standard for the identification of watermarks in historic papers, and are now in use at the leading German research libraries. Tight restrictions on eligible equipment costs in the funding scheme meant that we could not secure a thermographic set-up for this project. We remain extremely grateful to Herr Meinlschmidt for his advice at the time of our application.
25 Allan Stevenson, 'The Problem of the Blockbooks', in *Blockbücher des Mittelalters: Bilderfolgen als Lektüre. Gutenberg-Museum, Mainz, 22. Juni 1991 bis 1. September 1991* (Mainz: Philipp von Zabern, 1991), p. 243. Measurements (in mm): width 22, height 23 (the circle is not perfectly round), between chain lines of width 41/2. In the absence of high-resolution images and given the heavy colouration, we are not confident in distinguishing the twins of this pair.
26 Paul Needham, 'The Paper Stocks of Blockbooks. Allan Stevenson and beyond', in Bettina Wagner (ed.), *Blockbücher des 15. Jahrhunderts: Eine Experimentierphase im frühen Buchdruck*, Bibliothek und Wissenschaft, 46 (2013), p. 54.
27 The beta-radiographs are digitised at http://corsair.themorgan.org/vwebv/holdingsInfo?bibId=145322 [accessed 16 October 2024] with the watermarks printed in Needham, 'The Paper Stocks', p. 54.
28 Charles-Moïse Briquet, *Les Filigranes: dictionnaire historique des marques du papier dès leur apparition vers 1282 jusqu'en 1600*, 4 vols (Geneva: A. Jullien, 1907), no. 869.
29 J. S. G. Simmons and B. van Ginneken-van de Kasteele, eds, *Likhachev's Watermarks*, Monumenta chartae papyraceae historiam illustrantia 15 (Amsterdam: Paper Publications Society, 1994), nos 2603 ['Slavonic ms., 1460s'], 2612–13 ['Slavonic ms., mid-15th c.'] and 2960 ['Russian document, ?1450–62'], at pp. 238–40 with pll. 84–5.
30 PO-24768 to PO-24778, digitised in the *Wasserzeichen-Informationssystem*: https://www.wasserzeichen-online.de//wzis/index.php [accessed 16 October 2024].
31 Fréderic Ferber, 'La production et la commercialisation du papier à Metz et dans les pays messin à la fin du Moyen Âge', *Annales de l'Est* 57 (2007), 155–85, written apparently entirely without knowledge of Zaar-Görgens, *Champagne – Bar – Lothringen*, pp. 33–9, and Maria Zaar-Görgens, 'Papiermacherlandschaft Lothringen:

Zentren der Papierherstellung an Obermosel und Meurthe (ca. 1444–1600) unter besonderer Berücksichtigung der städtischen Papiermacherei in Metz', *Kurtrierisches Jahrbuch* 35 (1995), 167–88.

32 Ferber, 'La production', 159–64 and 169–74; Zaar-Görgens, *Champagne – Bar – Lothringen*, pp. 33–8.
33 Ferber, 'La production', 174–6; Zaar-Görgens, 'Papiermacherschaft Lothringen', 183–6.
34 Ferber, 'La production', 177; Zaar-Görgens, 'Papiermacherschaft Lothringen', 183–4 and 186.
35 Zaar-Görgens, *Champagne – Bar – Lothringen*, p. 69.
36 Ferber, 'La production', 176–7 and 184; Zaar-Görgens, *Champagne – Bar – Lothringen*, pp. 34–5.
37 Zaar-Görgens, 'Papiermacherlandschaft Lothringen', 181 n. 79.
38 Stevenson, 'The Problem of the Blockbooks', p. 243.
39 The proposition expressed by Elke Purpus that 16119-1 must have been printed in the mid-1460s on old paper that had been stored for some years is improbable. That explanation is only required if one accepts some quite uncertain previous watermark datings of copies of state IV C (which must necessarily precede IV D), and places too much weight on an assumption that 16119-1 must have been printed only shortly prior to its binding by Richenbach in 1467: see Purpus, *Die Blockbücher der Apokalypse*, pp. 166–9. What is more, all copies of state IV D must have been printed before any of state IV E, and Purpus accepts watermark datings of at least two of those to 1462/3.
40 Ad Stijnman, 'The Colours of Black: Printing Inks for Blockbooks', in Bettina Wagner (ed.), *Blockbücher des 15. Jahrhunderts: Eine Experimentierphase im frühen Buchdruck*, Bibliothek und Wissenschaft 46 (2013), 67–9.
41 Ira Rabin, 'Material Studies of Historic Inks: Transition from Carbon to Iron-Gall Inks', in Lucia Raggetti (ed.), *Traces of Ink: Experiences of Philology and Replication*, Nuncius Series 7 (Leiden: Brill, 2021), pp. 70–8; Claudia Colini, Ivan Shevchuk, Kyle Ann Huskin, Ira Rabin and Oliver Hahn, 'A New Standard Protocol for Identification of Writing Media', in Jörg B. Quenzer (ed.), *Exploring Written Artefacts: Objects, Methods and Concepts*, vol. 1, Studies in Manuscript Cultures 25 (Berlin/Boston: De Gruyter, 2021), 161–81. The authors are indebted to Stefan Hanß and Ira Rabin for their readiness to undertake this analysis, provision of the technical equipment and generosity of time and advice.
42 Rabin, 'Material Studies of Historic Inks', p. 73.
43 Purpus, *Die Blockbücher der Apokalypse*, pp. 170–3; the principal evidence is that provided by the dialect analysis of the numerous interleaved German translations, assembled and discussed by Nigel F. Palmer, 'Latein und Deutsch in den Blockbüchern', in Nikolaus Henkel and Nigel F. Palmer (eds), *Latein und Volkssprache im deutschen Mittelalter 1100–1500. Regensburger Colloquium 1988* (Tübingen: Niemeyer, 1992), pp. 316–25.
44 Palmer, 'Blockbooks', p. 11 (to BB-2).

The Trier Psalter-Hymnal (Manchester, John Rylands Research Institute and Library, MS Lat. 116): Palette and Pigments

RICHARD GAMESON, UNIVERSITY OF DURHAM
ANDREW BEEBY, UNIVERSITY OF DURHAM

Abstract

The findings of recent scientific investigation into the pigments employed in the ninth-century Psalter-Hymnal, JRRIL, MS lat. 116, are reported here. The pigments of this book are compared with those that have been reliably identified in other Carolingian manuscripts, with particular attention being paid to the use (or otherwise) of vermilion. Changing patterns in the utilisation of vermilion in manuscripts from late antiquity to the high Middle Ages are outlined, and possible reasons for fluctuations are considered.

Keywords: Carolingian manuscripts; psalter-hymnal; pigment identification; vermilion

The psalter-hymnal, JRRIL, MS lat. 116, is a grand volume that, even after brutal trimming, measures 425 × 320 mm (written area, 349 × 210 mm);[1] it was probably produced early in the third quarter of the ninth century.[2] The large format of the book along with the fact that it features a hymnal as well as a psalter indicate that it was designed for use by a religious community rather than an individual; and the many revisions and additions that it subsequently received – not to mention several replacement leaves – show that it had a long working life.[3] The additions also establish the manuscript's medieval provenance, since obits inserted into the calendar and a list of monks added to fol. 11r demonstrate that it belonged to the abbey of St-Maximin, Trier (Rhineland-Palatinate).[4] The hands responsible for most of these insertions were active in the last quarter of the eleventh century and the first quarter of the twelfth, however, more than 200 years after the book was made.[5] This period is reduced by half if one accepts the proposed localisation to Trier of the three hands responsible for adding, sequentially, a prayer, a hymn and a canticle to fol. 109r, probably during the second and third quarters of the tenth century;[6] yet there still remains a gap of about 100 years between the manufacture of the original manuscript and the earliest evidence for its presence in Trier. Views on whether the volume was made in Trier (be it at St-Maximin or at Trier Cathedral) or merely came to be owned there have differed, in part according to the evidence that was prioritised (the decoration of the volume may be compared with that of a fine gospel-book of Trier provenance, albeit itself of uncertain origin,[7] while the hand of the psalter-hymnal's original scribe is linked to Tours).[8] Any credible case for a local origin would have to explain how the manuscript managed to survive the sacking of Trier and the burning of St-Maximin by Vikings in 882.[9]

Although many aspects of MS Lat. 116 have received scholarly attention, its pigments have not hitherto been discussed, doubtless because it was not until relatively recently that appropriate methods for examining them became available.[10] Happily, the colourants in question can now be reliably identified via non-invasive techniques, and the present study reports the findings of a recent campaign that investigated the inks and pigments of the manuscript using multispectral imaging (Figures 1–4), reflectance spectroscopy, Raman spectroscopy (Figure 5), X-ray fluorescence spectroscopy and photomicroscopy (Figure 6). In addition, false-colour imaging was employed to render the placement of individual pigments readily apparent to the naked eye (Figures 3–4). Further details of the equipment that was used for the analyses are provided elsewhere.[11] First, we outline the decorative programme of the manuscript; then, we discuss the pigments and painting techniques that were utilised to realise it. Finally, we contextualise the colourants employed in MS lat. 116 by comparing them with those of other Carolingian books whose pigments have likewise been securely identified.

JRRIL MS lat. 116 is a colourful book. One prefatory text (fol. 8r: Figure 1),[12] the rubric to Psalm 1 (fol. 15v) and then the incipits of Psalms 1, 51 and 101 (fols 16r, 40r (Figure 2) and 64r) are all marked by large decorated initials rendered in red, orange, yellow, green and blue, plus black or brown; these letters are accompanied by display capitals of various forms written either in orange or in brown/black, and stroked in one or more of red, orange, yellow, green and blue. Smaller decorated letters in the same colours introduce a prologue (fol. 12r) and the lengthy Psalm 118 (74r).[13] All other psalms, canticles and hymns, along with the subsections of Psalm 118, are headed by three- to four-line-high initials outlined in black/brown, their bodies stroked with colour. Many of these letters were enhanced with foliate tufts, some with simple interlace and knotwork,[14] a few with bird heads and rosettes, all coloured,[15] while a couple are outlined in orange dots in the Insular manner.[16] Those heading psalms at the liturgical divisions, along with those for the first canticle and the first hymn, are slightly larger, more elaborate and more colourful than the rest.[17] Each psalm, canticle and hymn verse is marked by a one-line-high capital stroked in one or two colours. The titulus that precedes and the rubric for the prayer that follows each psalm are written in orange Uncials; the prayers themselves are headed by an orange initial. The 'KL's plus tilde (an abbreviation mark), collectively representing *Kalendae*, that introduces each page of the calendar is shaded in two or more of red, yellow, green and brown; the rubrics for each month and the numerals for each day therein are written in orange, with a long run of 'N's (for 'natalis'), one for almost every day, in orange stroked with green.[18]

With regard to the psalm and verse initials, the parity of ink tone and density between these letters and the script they accompany, along with the way that they

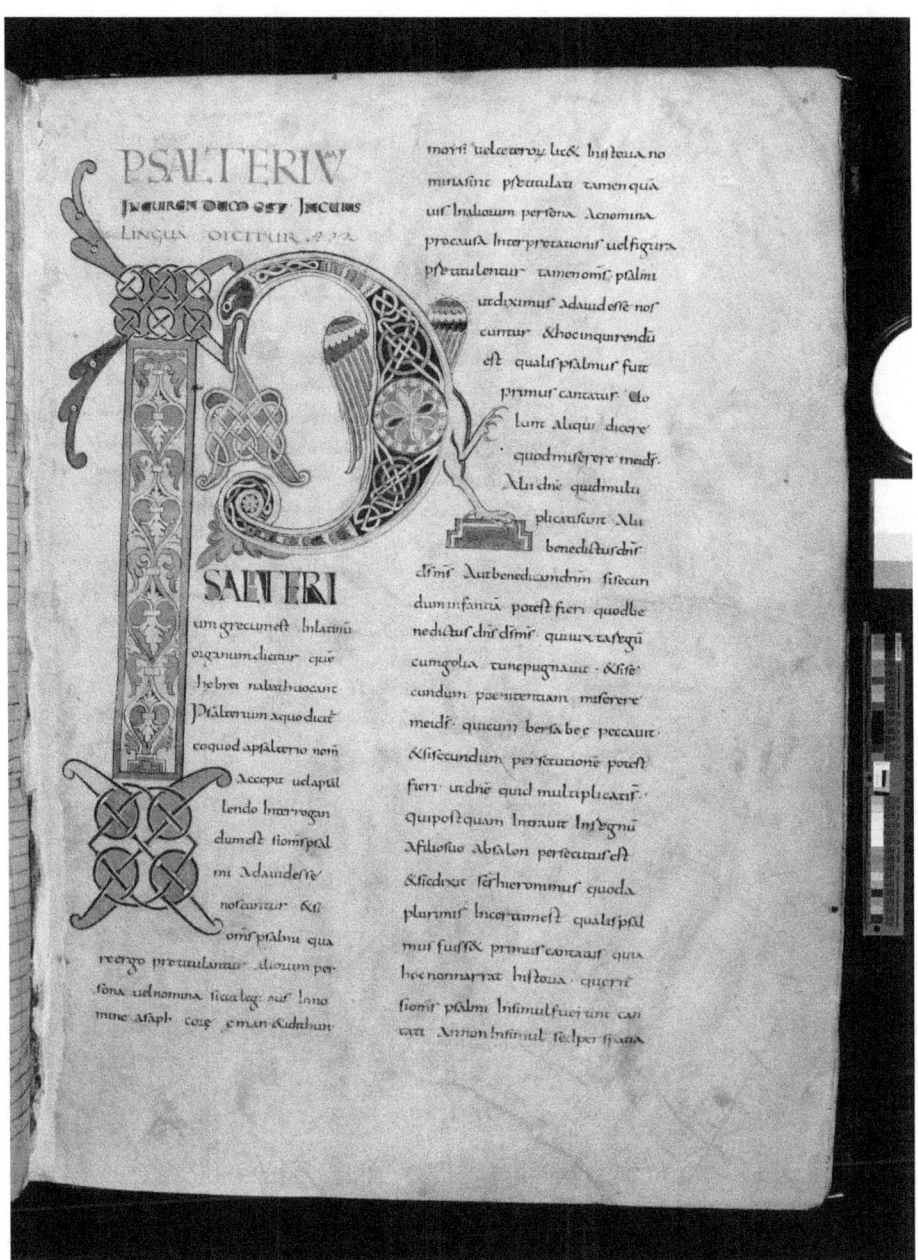

Figure 1 JRRIL, MS Lat. 116, fol. 8r (RGB image, reconstructed from multispectral imaging data cube). © The John Rylands Research Institute and Library, University of Manchester.

Figure 2 JRRIL, MS Lat. 116, fol. 40r (RGB image constructed from multispectral image data cube). © The John Rylands Research Institute and Library, University of Manchester.

are integrated into the lines of writing, strongly suggest that they were the work of the scribe of the main text. Whether the same hand was also responsible for some of the grander initials is more difficult to establish. He is unlikely to have executed all of them, since there are two distinct styles of work, implying the participation of two separate hands. The first two large initials (for a preface and for the rubric to Psalm 1, fols 8r (Figure 1) and 15v respectively) are better conceived and more confidently executed than the remaining three (marking the incipits to Psalms 1, 51 and 101; fols 16r, 40r (figure 2) and 64r); they are further distinguished from the latter by a more even application of paint, and by the use of a darker brown ink for key elements within the design. The circumstance that both these superior initials appear in the same quire (quire 2) and, moreover, on the same side of a single folio therein indicates that they would have been undertaken at the same time, and is wholly compatible with the hypothesis that they are the work of different artists.[19]

The pigments that were used to realise the palette of the Trier Psalter-Hymnal were as follows (for specimens of the spectra that permitted these identifications see Figure 5):

Red: vermilion (highlighted in Figure 3)
Orange: red lead (highlighted in Figure 3); red lead + vermilion; red lead + organic
Yellow: orpiment
Green: copper (verdigris) (highlighted in Figure 3)
Blue: indigo; indigo + verdigris (highlighted in Figure 4)
Brown (light and dark): iron gall without vitriol (highlighted in Figure 3).[20]

In addition, plain parchment was incorporated into the designs as a neutral tone or 'white'.[21]

The colours were applied and presented individually, generally being isolated one from another by brown or black lines and/or plain parchment, following the manner of Insular illuminators. In the less meticulous work of the second artist, small areas of ink and colour do occasionally touch, resulting in modest contamination and points of discoloration. All the colour was applied with pens rather than brushes; this was done more evenly in the first two initials than in the other three: in the latter, although certain areas of a given colour are dense and dark, others are thin and 'scratchy' with visible pen strokes. Additional problems were caused by the copper green that was employed in the first quire of the psalter text (fols 16–23); this was evidently a particularly aggressive formulation that has severely damaged – to the extent of erosion – much of the parchment to which it was applied (elsewhere in the manuscript, the greens have penetrated but not perforated the parchment).

Very few Carolingian manuscripts have had their inks and pigments identified by modern scientific techniques, and a relatively high proportion of those for which there is reliable information are atypical in one way or another (almost a fifth of the total, for example, comprise manuscripts associated with Charlemagne). Conversely, the thirty-five volumes for which data are currently available represent the work of at least twelve and possibly sixteen different centres (Arras, ?Dol,[22] Echternach, ?Fleury, Fulda, Hautvillers, Liège, Lorsch, ?Mainz, ?Metz, Orléans, Reims, Saint-Amand, Salzburg and Tours, as well as the 'court schools' of Charlemagne), range in date from the late eighth century to end of the ninth and include a couple of undecorated volumes as well as several extensively decorated ones.[23] Thus, while it is currently impossible to discuss features at the level of individual scriptoria or even of regions, it is feasible to distinguish between materials that were relatively common and apparently widespread and those that were not.

If a pigment that was used in at least half of the Carolingian manuscripts to have the colour in question may be defined as 'common', then all but one of the colourants employed in the Trier Psalter-Hymnal were common ones, albeit to different degrees. At the upper extreme of popularity are red lead, indigo and orpiment, since

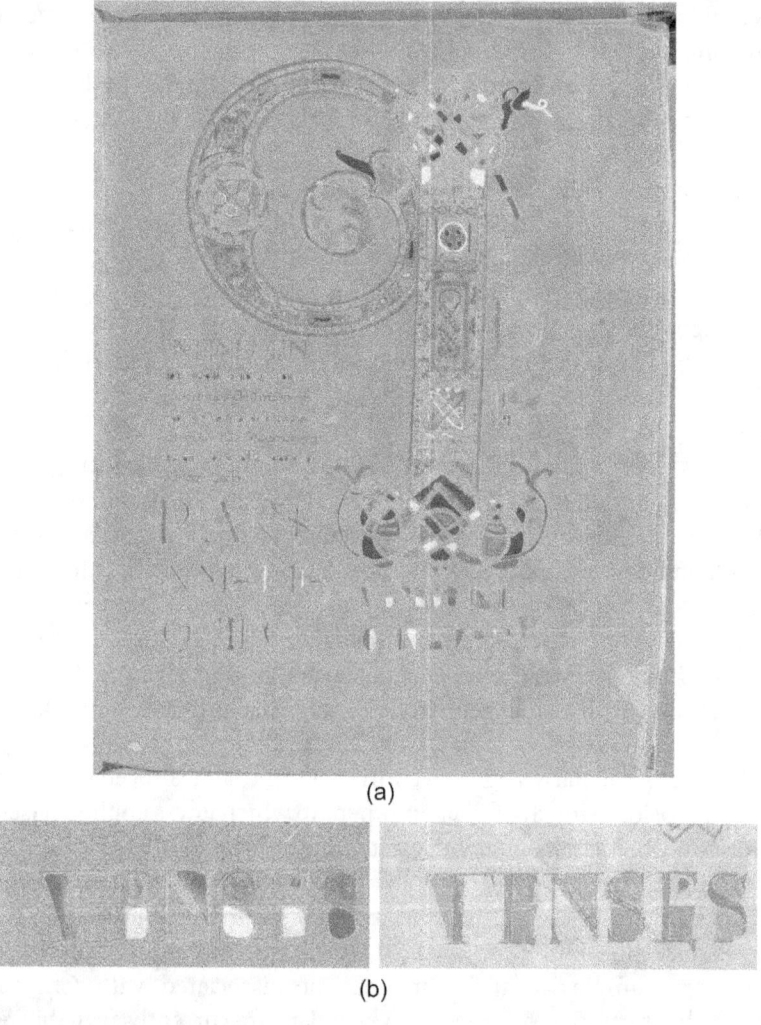

Figure 3 (a) JRRIL, MS Lat. 116, fol. 40r: pigment classification map created by principal component analysis of multispectral imaging data. PCs 1, 3 and 7 selected to show red lead (red), vermilion (yellow) and copper green (blue). © The John Rylands Research Institute and Library, University of Manchester. (b) Detail of display script on fol. 40r: (left) RBG image, and (right) PCA image, highlighting red lead (red), vermilion (yellow) and copper green (blue); gallo-tannic ink appears as green. © The John Rylands Research Institute and Library, University of Manchester.

Figure 4 JRRIL, MS Lat. 116, fol. 40r: pigment classification map creating by principal component analysis of multispectral imaging data; PC's 1, 2 and 4 selected to show indigo (blue), read lead (magenta) and copper green (yellow). © The John Rylands Research Institute and Library, University of Manchester.

every manuscript with orange, blue and yellow (twenty-seven, twenty-five and twenty-four out of thirty-five volumes respectively) features these materials. Accordingly, they may be described as the standard orange, blue and yellow. Nevertheless, a distinction can be drawn between red lead on the one hand and orpiment and indigo on the other: whereas the former was the sole pigment utilised for orange (albeit sometimes modulated with other substances), the latter pair were not the only materials employed for yellow and blue. Orpiment and indigo were the

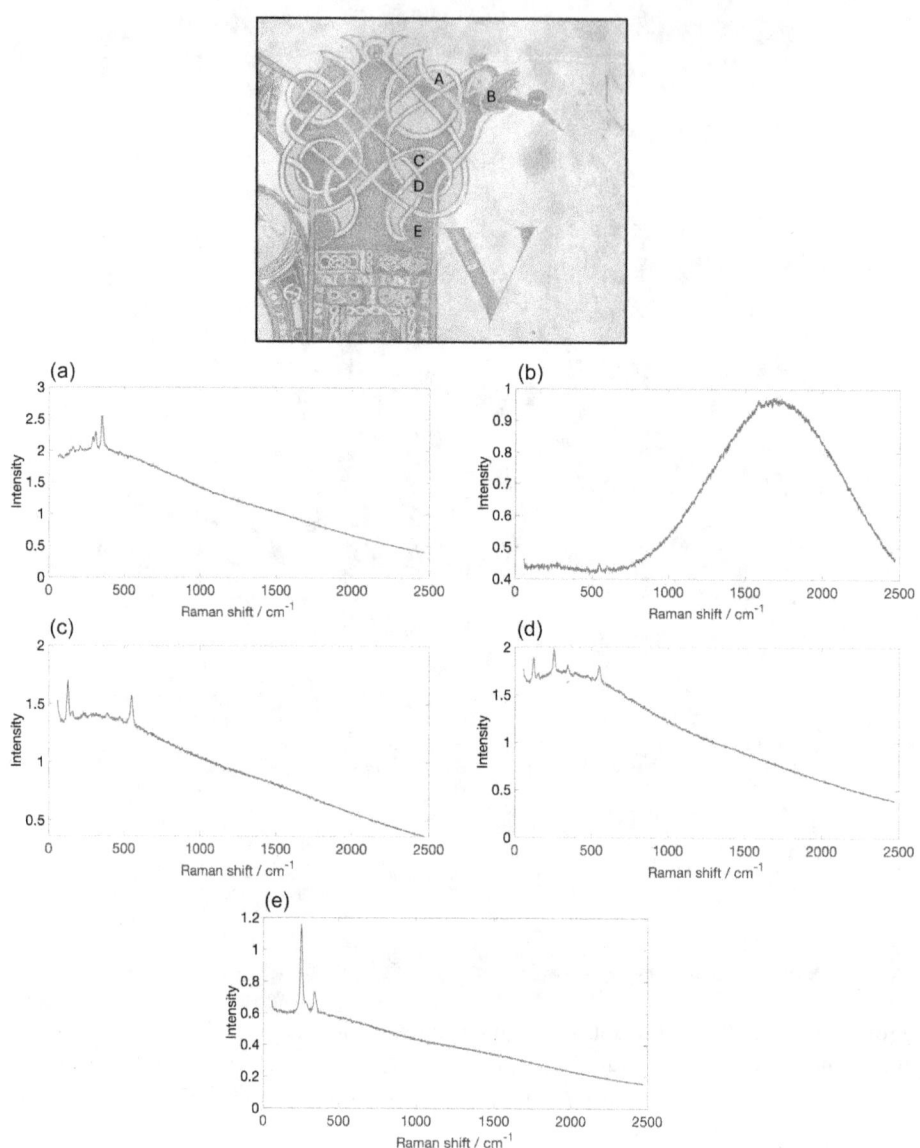

Figure 5 Raman spectra from selected areas of JRRIL, lat. 116, fol. 40r, showing the presence of (a) orpiment (yellow), (b) indigo (blue), (c) red lead (orange), (d) red-lead with traces of vermilion (orange) and (e) vermilion (red). © The John Rylands Research Institute and Library, University of Manchester.

sole yellow and blue colourants used in the Trier Psalter-Hymnal, but yellow ochres were deployed alongside orpiment in six other Carolingian volumes, while lapis lazuli and/or azurite appear alongside indigo in thirteen books.

In relation to green, the situation is a little more complicated. Fewer volumes of the small corpus feature the colour (twenty-one out of thirty-five) than have orange, yellow or blue, yet there is greater diversity in the pigment used to create it. Whereas in the Trier Psalter-Hymnal, all the greens were verdigris (a range of substances, depending on the nature of the copper salts and the minutiae of the eventual paint mixture),[24] six Carolingian manuscripts have only vergaut greens (created by combining the woad blue with the orpiment yellow), while seven boast both verdigris and vergauts. In one, a green earth was also employed; in another, malachite too;[25] and in yet another, indigo served as green.[26]

The picture is most diverse in relation to red, which – thanks to its employment in rubrics as well as for artwork – appears in all but one out of thirty-five volumes (and in the sole exception, orange fulfils this role).[27] This is the one case where the colourant employed in the Trier Psalter-Hymnal fails to qualify as 'common' according to our self-imposed criterion (albeit only just): sixteen out of thirty-four volumes with red feature vermilion, while eighteen do not. The creators of the latter volumes resorted instead to ochres or red lead (with or without the admixture of an organic colorant). Of the sixteen books wherein vermilion was employed, eight boast one or more other colourant(s) for red alongside, or in mixtures with, vermillion.

Vermilion can be obtained from the mineral cinnabar, a form of mercury (II) sulfide (HgS).[28] Alternatively, it may be manufactured by grinding together and then heating mercury and sulfur. Unfortunately, it is impossible to distinguish between the two forms using non-invasive analytical techniques. Vermilion appears in some late-antique books,[29] but whether it continued to be employed in early medieval ones is currently unclear (it is certainly absent from Insular manuscripts).[30] The limited evidence of our thirty-five volumes suggests a relatively widespread if episodic employment of vermilion in Carolingian scriptoria, since it appears in manuscripts from Arras, Fulda, Hautvillers, Liège, Lorsch, Reims, Saint-Amand, Salzburg and Tours, as well as from the 'court schools' of Charlemagne. Conversely, the pigment does not seem to have been used on the same scale in the tenth century; it was not part of the palettes of most Ottonian illuminators.[31] Usage of vermilion revived during the course of the eleventh century (the point at which it was first adopted in Britain),[32] and by the end of the century it had become the standard colourant for red in Western books, a position that it retained for the rest of the medieval period.

The reasons for these changing patterns of usage are obscure, but they may be related to fluctuations first in the availability of cinnabar and then (putatively) in the manufacturing of vermilion. Most of the cinnabar of antiquity came from the mines at Almadén, in central southern Spain (whether vermilion was synthesised in the ancient world and used alongside cinnabar is an open question). Pliny the Elder (d. 79 AD), the principal literary source for pigments and their use in antiquity, highlighted the importance of the mines at Almadén, noted that cinnabar red was sometimes adulterated with red lead and included 'in the writing of books' among the uses to which it was put.[33] The output of these mines may have declined in late antiquity.[34] Be that as it may, the Islamic conquest of Spain in 711 and the circumstance that the cinnabar from Almadén was increasingly employed in the creation of mercury under the new regime must have affected the availability of the mineral in the early medieval West (the region around Almadén was not reclaimed by Christian forces until the mid-twelfth century).[35] The earliest extant Latin recipe for manufacturing vermilion appears in the so-called *Compositiones variae* or *Compositiones Lucenses*.[36] Although the contents of this collection are believed to date back to late antiquity, the single known manuscript copy of the work was made *c*. 800, precisely the period when vermilion reappears in some quantity in European books.[37] Whether dwindling need for the colourant, a loss of knowledge of how to synthesise it or a decrease in the supply of mercury was the principal factor behind the reduced presence of vermilion in northern European books during the tenth century is currently unclear. So too, reciprocally, is whether it was primarily a resurgence in demand for the colourant, a rediscovery of the relevant alchemical knowledge or an enhanced supply of mercury that led to the revival of vermilion's use during the eleventh century. It is to be hoped that, as future work adds more detail to this picture, the relative importance of these (and other) factors may become clearer in both cases.

The rationale for the restricted palette of the Trier Psalter is not hard to divine: the Carolingian manuscripts with more expansive ranges of pigments are typically those with illustrations, whose creators required not only more colours but also different shades of the same hue to achieve the desired painterly effects. A smaller range of unmodulated colours, by contrast, was well suited to the realisation of bold, decorated initials, as had been demonstrated by Insular and Merovingian practitioners in earlier centuries. Just as the ornamental vocabularies of these seventh- and eighth-century scribe-artists – interlace, knotwork, bird and beast heads, stylised foliate tufts and marigold motifs – lay behind that deployed in the Trier Psalter, so too did their restricted palette. The pigments were also the same, with the exception (at least in relation to Insular books) of vermilion.

In terms of quality, the main artist of the Trier Psalter-Hymnal (the hand that accomplished the initials for Psalms 1, 51 and 101 (fols 16r, 40r and 64r)) lacked the finesse of the greatest Insular practitioners, responsible for the superlative decoration of such manuscripts as the Durham Gospels, the Lindisfarne Gospels and the Book of Kells.[38] Although the interlace and knotwork designs in the Psalter-Hymnal are relatively simple, they still include awkward passages that betray less than perfect control of the craft.[39] In the 'Q' on fol. 40r, for example, the interlace is rubbery, certain elements that were manifestly supposed to be symmetrical do not in fact match and forms that would ideally have been circular or semi-circular are eliptoid or otherwise distorted (Figures 2 and 6); while in the 'D' on fol. 64r the interlace within a couple of panels at the top right of the letter-curve went awry. Similar deficiencies in design appear in the psalm initials: the knotwork twists in

Figure 6 Micrographs of two areas (10 × 15 mm) of JRRIL, lat. 116, fol. 40r: the upper panel features copper-green, red lead, orpiment and gallo-tannic ink, the lower one features copper-green and gallo-tannic ink. © The John Rylands Research Institute and Library, University of Manchester.

the initial for Psalm 17 (22r), for example, are irregular, and the alternation of colour between green and 'white' breaks down at one point; the knotwork for Psalm 85 (57r), though simple, is likewise irregular; the supposedly symmetrical interlace terminals of the 'B' for Psalm 118 (74r) are distorted and off-true; and the motif of a pair of strands threading through a circle that was used twice in the 'D' for Psalm 7 (18r), thrice in that for Psalm 52 (40v), as also once in the 'C' for the canticle of Isaiah (88v) and in the 'P' heading a hymn on 95r, is wobbly and unbalanced, the circles themselves more or less distorted.[40]

Imperfect realisation of interlace design is far from unique to the Trier Psalter-Hymnal. Even in the Insular world, there are few manuscripts that can match the perfection of Durham, Lindisfarne and Kells in this regard: the mid-eighth-century Kentish Codex Aureus is unquestionably the most opulent book to survive from early England, yet the interlace in its canon tables lacks the fluidity and grace of that in those manuscripts.[41] Flawed interlace is a feature of many ninth-century continental manuscripts. This is most obviously the case in relation to books from Brittany but is also true – albeit to a lesser extent – of examples from other locations, not least Tours (the centre with which the original scribe of the Trier Psalter-Hymnal has been linked).[42] And imperfect interlace even appears in some of the most luxurious volumes of the age, including the Godescalc gospel-lectionary, commissioned by Charlemagne between 781 and 783,[43] a gospel book made for Ebo, archbishop of Reims (*sedit* 816–36 and 840–1),[44] and a gospel book that was produced in Tours between 849 and 851 for Emperor Lothar.[45]

Correspondingly, the way in which the colours were added to the Trier Psalter was less than optimal. Both artists applied the coloured inks with a pen (as was the norm), but while the first achieved a relatively even coverage, the second did not: the larger areas of colour within his letters tend to be slightly blotchy, the individual pen strokes more or less visible (Figure 6). Equally, while allowance must be made for subsequent discoloration (especially of the reds and oranges), some of the hues are a little subdued. Perhaps because supplies of the mineral itself were limited, the orpiment yellow in particular is intermittently highly dilute and hence pale. The result remains eye-catching; however, had only modest extra attention been paid to the preparation of the pigments and to how they were applied, it could have been even more so.

To sum up: the palette and pigments of the Trier Psalter-Hymnal are – in so far as the limited comparanda currently available permit one to judge – fairly typical of a Carolingian book with major but non-figural decoration. The coloured inks were prepared to an acceptable rather than to a high standard. They were seemingly applied by two hands, one rather more skilled and careful than the other, with results that are dramatic but lack finesse. Given the dearth of reliable information about the pigments and paintings techniques of most Carolingian manuscripts, these features are currently of little help in relation to the question of where the Trier Psalter-Hymnal was made; as further manuscripts are investigated with appropriate techniques, however, the facts recorded here may permit it to be contextualised more fully and accurately. Be that as it may, future work may reasonably be expected to bring clarification to the intriguing history of vermilion in the early Middle Ages and, by extension, to enhance understanding not only of the trade in mercury but also of the spread of alchemical knowledge for processing and using it.

Notes

1 As attested by the narrow upper margin, the cropping of the decorated initial at the top of fol. 74r, and the loss of almost all the prickings (exceptionally, prickings are preserved in the outer margin of fol. 23). The parchment, once stout, is weathered, worn and discoloured from extensive use, with erosion particularly pronounced at the lower outer corners. There is minimal contrast between hair and flesh sides. It appears to have been arranged HF, FH. Ruling was applied to hair sides, two sheets at a time.

2 The manuscript cannot have been made before c. 850, since its calendar includes an obit for Thegan of Trier (who died between 848/49 and 853) written by the original scribe, while the nature of its script and decoration indicate that it is unlikely to have been produced much after this date.
3 The fifty-seven hymns (fols 95r–104v) are followed by eighteen Old Testament canticles (fols 104v–109r). The hymns are indexed in J. Mearns, *Early Latin Hymnaries. An Index of Hymns in Hymnaries before 1100* (Cambridge: Cambridge University Press, 1913), siglum 'GB', the MS cited by its former shelf-mark, 'Crawford MS lat. 133'.
4 The obits are published in F. Roberg (ed.), *Das älteste 'Necrolog' des Klosters St. Maximin vor Trier*, Monumenta Germaniae Historica, Libri Memoriales et Necrologia, nova series 8 (Hannover: Hahnsche, 2008). The list of monks (from the time of Ogo I, abbot of St-Maximin 934–45) was edited by O. Holder-Egger as 'Nomina monachorum S. Maximini Treverensis', in G. Waitz (gen. ed.), *Monumenta Germaniae Historica Scriptores XIII* (Hanover: Hahnsche, 1881), pp. 301–2.
5 Hands 'H_6' and 'H_7' according to the classification of Roberg (ed.), 'Necrolog', pp. 19–56.
6 All three linked to Trier by H. Hoffmann, *Buchkunst und Königtum im ottonischen und frühsalischen Reich*, 2 vols, Monumenta Germaniae Historica Schriften 30, 1 (Stuttgart: Anton Hiersemann, 1986), p. 480.
7 Trier, Stadtbibliothek, 23/122a/b 2°: M. Embach, *Hundert Highlights. Kostbare Handschriften und Drucke der Stadtbibliothek Trier* (Regensburg: Schnell & Steiner, 2013), no. 5; L. Nees, *Frankish Manuscripts. The Seventh to the Tenth Century*, 2 vols (London: Harvey Miller – Turnhout: Brepols, 2022), no. 39.
8 For a summary of previous views on its origin, see Roberg (ed.), *Necrolog*, pp. 4–6, who himself leaves the matter open, concluding that 'Die Handschrift wird in Tours oder St. Maximin geschrieben worden sein'. On the early library of St-Maximin more generally, see I. Knoblich, *Die Bibliothek des Klosters St. Maximin bei Trier bis zum 12. Jahrhundert* (Trier: WVT Wissenschaftlicher Verlag, 1996); she favours production of MS Lat. 116 at Tours, and has the book reaching Trier in the aftermath of 882 (pp. 58–9, 84–5).
9 'hoc monasterium ... non solum concrematum est ...': Sigehard of St-Maximin, *Miracula Maximini*, ch. 3: G. Henschen and D. Papebroch (eds), *Acta Sanctorum, Maius*, 7 (Paris, 1867), p. 31. Sigehard was writing in the 960s.
10 Key published descriptions and studies include M. R. James, *Descriptive Catalogue of the Latin Manuscripts in the John Rylands Library at Manchester, Volume I*, repr. with addenda and corrigenda by F. Taylor (Munich: Kraus, 1980), pp. 211–7; Hoffmann, *Buchkunst und Königtum*, 1, p. 480; B. Bischoff, *Katalog der festländischen Handschriften des neunten Jahrhunderts*, ed. Birgit Ebersperger, 2 (Wiesbaden: Harrassowitz, 2004), p. 170 (no. 2679); and Roberg (ed.), 'Necrolog', pp. 12–18. For the calendar, see F. Roberg, 'Der sogenannte Lorscher Prototyp und der Kalender Manchester, John Rylands Library lat. 116. Beobachtungen zur Entwicklung der Gattung Kalender: mit einem Editionsanhang', *Archiv für Diplomatik, Schriftgeschichte, Siegel- und Wappenkunde* 53 (2007), 27–58. On the now incomplete litany (fols 112r–113r), see M. Coens, 'Anciennes litanies des saints (suite)', *Analecta Bollandiana*, 55 (1937), 49–69.

11 The XRF analysis was done by Ira Rabin of the Bundesanstalt für Materialforschung und -prüfung, Berlin, whom we thank most warmly for kindly sharing her data with us. All the other analyses were accomplished by the present writers. For explanation of these techniques and of their application to manuscripts, see R. Gameson, A. Beeby et al., *The Pigments of British Medieval Illuminators: A Scientific and Cultural Study* (London: Archetype, 2023), chapter 1 and appendix 3 (detailing the equipment).

12 '*Psalterium Inquirendum est*. Psalterium grecum est. In latinum organum dicitur ...': F. Stegmüller, *Repertorium Biblicum Medii Aevii*, 1 (Madrid: Consejo Superior de Investigaciones Científicas, 1950), no. 426.

13 'Primus psalmus ad Christi pertinet personam ...' (a catena of *tituli psalmorum*): Stegmüller, *Repertorium*, I, no. 416.

14 For example, both those on fol. 77v and one of those on fol. 78r.

15 Bird and beast heads appear on fols 18r, 20r, 44v, 46v, 56v, 74r, 80r, 92r, 93r and 105r; rosettes appear on fols 23v, 24v, 26v and 84r.

16 On fols 21r (Psalm 15), 21v (Ps. 16), 83v (Ps. 136) and 85r (Ps. 142).

17 Psalms 26 (fol. 26v), 38 (fol. 33v), 52 (fol. 40v), 68 (fol. 47v) [Psalm 80 (fol. 55r) is part of a bifolium (fols 48 and 55) that was replaced *s*. xii], 97 (fol. 62v) and 109 (fol. 71r); the Canticle of Isaiah (fol. 88v); and the hymn 'Primo dierum omnium' (fol. 95r).

18 All reproduced in (unnumbered) plates at the end of Roberg (ed.), *Necrolog*.

19 Fols 8 and 15 comprise a bifolium.

20 As also the ink of the text. XRF analysis by Ira Rabin confirmed that the ink (for text and artwork alike) was made without vitriol. On the inks of Carolingian manuscripts, see Z. Cohen, O. Hahn and I. Rabin, 'Black Carolingian Inks under Examination. Part 1: Recipes and Literature Survey', *Techne*, 55 (2023), 90–5; and Z. Cohen, T. Hennings, O. Hahn, P. Depreux and I. Rabin, 'Black Inks under Examination. Part 2: A Case Study of Carolingian Manuscripts', *Techne*, 56 (2023), 105–13.

21 It is conceivable that the thin strip of interlace within the upright of the 'E' in the display script on fol. 16r has been touched with white paint (if so, presumably white lead); however, it proved impossible to obtain useful measurements from this meandering line less than 1 mm wide.

22 If not specifically Dol, then somewhere else in Brittany.

23 Abbeville, Bibliothèque municipale, MS 4; Cambridge, Fitzwilliam Museum, MS 45-1980; London, British Library, MS Add. 11848; Manchester, John Rylands Research Institute and Library, MSS lat. 9, lat. 10, lat. 12, lat. 116 and lat. 174; Orléans, Médiathèque, MSS 17 and 295; Oxford, Bodleian Library, MSS Bodley 218, Bodley 579, Douce 59, Laud lat. 92, Laud lat. 102, Laud misc. 134 and Laud misc. 148; Paris, Bibliothèque de l'Arsenal, MS 599; Paris, Bibliothèque nationale de France, MSS lat. 2422, lat. 2423, lat. 5763, lat. 8849, lat. 8850, lat. 9380, lat. 9383, lat. 9387, lat. 9433, lat. 11937 and n.a.l. 1203; Stockholm, Kungliga Biblioteket, MS A.136; Trier, Stadtbibliothek, MS 23/122a/b 2°, and Inc. 126 2o binding, etc. (membra disjecta from a Touronian Bible); Utrecht, Bibliotheek der Rijksuniversiteit, MSS 32 and 163; Vienna, Kunsthistorisches Museum, Schatzkammer, MS Inv. XIII.18; Vienna, Österreichische Nationalbibliothek, MSS cod. 1861. The manuscripts in Manchester, Oxford and Utrecht were analysed by the present writers. Data on the others – full or

partial – are published in: R. Clark and J. van der Weerd, 'Identification of Pigments and Gemstones on the Tours Gospel: the early 9th century Carolingian Palette', *Journal of Raman Spectroscopy*, 35.4 (2004), 279–83; C. Denoël, P. R. Puyo, A.-M. Brunet, and N. Poulain Siloe, 'Illuminating the Carolingian Era: new discoveries as a result of scientific analysis', *Heritage Science* 6 (2018), article 28; M. Éveno, 'Étude des enluminures d'un manuscrit de Saint-Amand du IXe siècle par PIXE, spectrophotocolorimétrie et diffraction de rayons X', *ihrt.hypotheses.org/469* (2015); 'La fabrique de l'art. Couleurs et matériaux de l'enluminure' (online archive of analyses of BnF MSS: https://fabriques.inha.fr/s/fva/page/accueil); R. Fuchs, D. Oltrogge and O. Hahn, 'Farbmittel und Maltechnik der Bibel von St. Maximin', in R. Nolden (ed.), *Die Touronische Bibel der Abtei St. Maximin vor Trier* (Trier: Geselschaft für nützliche Forschungen, 2002), pp. 239–42; S. Haag and F. Kirchweger (eds), *Das Krönungsevangeliar des Heiligen Römischen Reiches* (Vienna: Kunsthistorisches Museum, 2014), esp. pp. 135–49; D. Jembrih-Simbürger, W. Vetter, C. Hofmann, M. Aceto and T. Rainer, 'The Dagulf Psalter (Austrian National Library, Cod. 1861): A Multi-Analytical Approach to Study Inks, Dyes and Pigments of This Early Carolingian MS', *Restaurator. International Journal for the Preservation of Library and Archival Material*, 45:2-3 (2024), 173–90; S. Panayotova, P. Ricciardi and M. Crippa, 'Breton Gospels', in S. Panayotova (ed.), *The Art and Science of Illuminated Manuscripts: A Handbook* (London: Harvey Miller; Turnhout: Brepols, 2020), pp. 181–6; P. Roger, 'Étude technique sur les décors de manuscrits carolingiens', in J.-P. Caillet and M.-P. Laffitte (eds), *Les Manuscrits carolingiens. Actes du colloque de Paris, Bibliothèque nationale de France, le 4 mai 2007*, Bibliologia 27 (Turnhout: Brepols, 2009), pp. 203–16; P. Roger and A. Bosc, 'Étude sur les couleurs employées dans des manuscrits datés du VIIIe au XIIe siècle provenant de l'abbaye de Fleury', in A. Dufour and G. Labory (eds), *Abbon, un abbé de l'an mil* (Turnhout: Brepols, 2008), pp. 414–36. Unfortunately, we have not been able to consult Thomas Rainer, 'Farbstoffe, Pigmente und Metalltuschen: Die Farbigkeit des Goldenen Psalters im Licht materialanalytischer Untersuchungen', in D. Ganz (ed.), *Der Goldene Psalter von St. Gallen, Cod. Sang. 22, St. Gallen Stiftsbibliothek. Kommentar zur Faksimile-Edition* (Luzern, 2024), pp. 149–57.

24 As noted above, at least two batches of slightly different composition were employed in Lat. 116, as reflected in the circumstances that the green used between fols 17 and 23 has regularly eroded the parchment, whereas that employed thereafter has only penetrated it; furthermore, the tone of the former is (now) appreciably darker than that of the latter.

25 Verdigris, vergaut and malachite are all reported for the gospel-book, BnF, MS lat. 8849 (Salzburg; *s.* ix^1).

26 Jembrih-Simbürger et al., 'Dagulf Psalter' do not report on the greens in that manuscript, which has therefore had to be left out of our survey (and calculations) here.

27 Namely Cambridge, Fitzwilliam Museum, MS 45-1980: Panayotova et al., 'Breton Gospels'; also F. Wormald and J. Alexander, *An Early Breton Gospel Book*, (Cambridge: Cambridge University Press for the Roxburghe Club, 1977).

28 The practicalities of extracting and processing the pigment were described late in the first century BC by Vitruvius: *De architectura libri decem*, Book VII, chapters 8–9.

29 Occurring, for example, in the late fourth- or early fifth-century Codex Bezae (Cambridge University Library, MS Nn.2.41; analysis by the present writers), the

sixth-century Codex Purpureus Rossanensis (Rossano, Duomo, s.n.: M. L. Sebastiani and P. Cavalieri (eds), *Codex Purpureus Rossanensis. Un codice e i suoi segreti* (Rome, 2020), esp. p. 87 (analysis by MoLAB)); and – confined to one quire – in the late sixth-century Gospels of Augustine of Canterbury (Cambridge, Corpus Christi College, MS 286: R. Gameson, A. Beeby and C. Nicholson, 'The Gospels of Augustine of Canterbury', in Panayotova (ed.), *Art and Science of Illuminated Manuscripts*, pp. 172–8). For the use of cinnabar in antiquity to tint epigraphic inscriptions, see most recently A. Coccato, G. Barone, P. Mazzoleni and J. Prag, 'Initial Investigations of Rubricated Inscriptions from Roman Sicily: Comparing the Material Evidence with Ancient Writers' Ideals', *Techne*, 57 (2024), 39–47.

30   See Gameson et al., *Pigments of British Medieval Illuminators*, chapter 2.
31   See, for example., D. Oltrogge and R. Fuchs, *Die Maltechnik der Codex Aureus aus Echternach. Ein Meisterwerk im Wandel* (Nürnberg: Verlag der Germanischen Nationalmuseums, 2009), pp. 153–67; R. Fuchs, 'Farbmaterialien und Maltechnik der Hildesheimer Bernwardhandschriften', in M. Müller (ed.), *Schätze im Himmel, Bücher auf Erden. Mittelalterliche Handschriften aus Hildesheim* (Wiesbaden: Harrassowitz, 2010), pp. 161–92; R. Fuchs and D. Oltrogge, 'Das Werk der Maler. Zu Maltechnik und Materlialität der Miniaturen', in B. Schneidmüller, H. Walter-von dem Knesebeck, B. Wagner, R. Fuchs and D. Oltrogge, *Die Bamberger Apokalypse. Visionen vom Ende der Zeit* (Darmstadt: Wissenschaftliche Buchgesellschaft, 2022), pp. 365–80; and D. Oltrogge and R. Fuchs, 'Materielle Pracht und Kunstlerische Praxis: Kunsttechnologische Befunde an Handschriften der Liuthar-Gruppe', in W. Zimmerman, O. Siart, and M. Gedigk (eds), *Die Klosterinsel Reichenau im Mittelalter. Geschichte, Kunst, Architektur* (Regensberg: Schnell & Steiner, 2024), pp. 256–71.
32   Gameson et al., *Pigments*, p. 112.
33   'in voluminum ... scriptura': *Historia naturalis* XXXIII.39–40, §§ 117–122: Pliny, *Natural History* vol. IX, Books XXXIII–XXXV, ed. and trans. H. Rackham (Cambridge, MA: Harvard University Press; London: William Heinemann, 1952), pp. 88–92. Mining there finally ceased in 2002.
34   J. Edmondson, 'Mining in the late Roman Empire and beyond: Continuity or Disruption?', *Journal of Roman Studies*, 79 (1989), 84–102, although focusing on Iberia, says nothing about Almadén; however, since most of the general factors that he adduces to support a hypothesis of declining output in other mines would have applied equally there, it seems likely that this would have been the case for cinnabar production, too.
35   For an overview of mercury production there, albeit focusing on post-medieval developments, see J. Tejero-Mazanares, I. Garrido Sáenz, F. Mata Cabrera and M. L. Rubio Mesas, 'La metalurgia del mercurio en Almadén: desde los hornos de aludeles a los hornos Pacific', *Revista de Metalurgia*, 50:4 (2014), https://doi.org/10.3989/revmetalm.033.
36   *A Classical Technology Edited from the Codex Lucensis 490*, ed. John Burnam (Boston: The Gorham Press, 1920), pp. 44–5 and 70. On the manuscript tradition, nature and classification of this collection more generally, with references to further literature, see G. Frison and G. Brun, '*Compositiones Lucenses* and *Mappae Clavicula*: two traditions or one? New Evidence from Empirical Analysis and Assessment of the Literature', *Heritage Science*, 6 (2018), article 24. For modern experimentation with medieval

recipes for vermilion see M. J. Melo and C. Miguel, 'The Making of Vermilion in Medieval Europe: Historically Accurate Reconstructions from *The Book on How to make Colours*', *Fatto d'Archimia. Los pigmentos artificiales en las técnicas pictóricas* (Madrid: Ministerio de Educación Cultura y Deporte, 2012), pp. 181–95; and C. Miguel, J. V. Pinto, M. Clark and M. J. Melo, 'The Alchemy of Red Mercury Sulphide: The Production of Vermilion for Medieval Art', *Dyes & Pigments*, 102 (2014), 210–17.

37 Lucca, Biblioteca Capitolare Feliniana, MS 490: Bischoff, *Katalog*, II, no. 2524.

38 Respectively Durham Cathedral Library, MS A.II.17; London, British Library, MS Cotton Nero D.iv; and Dublin, MS Trinity College, 58.

39 The most accomplished work is generally the single-strand (single pen-line) flourishes of interlace that were used as terminals such as those on fols 77v, 78r, 80r, 80v and 83v.

40 The renderings of the motif in the 'D's for Psalms 89 (fol. 59v) and 109 (fol. 71r) are better, let down only by disjunctions in the circle itself. Imperfections are not limited to the Insular decorative repertoire, however: the four 'petals' within the 'O' for the hymn *O lux beata Trinitas* (fol. 98v) are uneven and asymmetrical, as are the eight of the similar motif within the 'D' of Psalm 26 (fol. 26v).

41 Stockholm, Kungliga Biblioteket, MS A.135. Facsimile: R. Gameson (ed.), *The Codex Aureus: An Eighth-Century Gospel Book*, 2 vols, Early English Manuscripts in Facsimile 28–9 (Copenhagen: Rosenkilde & Bagger, 2001–2).

42 For reproductions of relevant pages from Breton books see Wormald and Alexander, *Early Breton Gospel Book*, plates 33–4. For examples from Tours, see E. K. Rand, *A Survey of the Manuscripts of Tours*, Studies in the Script of Tours I, 2 (Cambridge, MA: Mediaeval Academy of America, 1929), plates 43, 46, 50, 54, 63, 68, 76, 95, 132, 136 and 154.

43 Paris, Bibliothèque nationale, MS n.a.l. 1203: W. Koehler, *Die Hofschule Karls des Grossen*, Die karolingischen Miniaturen 2, 2 vols (Berlin: Deutscher Verein für Kunstwissenschaft, 1958), 1, pp. 22–8, and 2, plates 1–12; M.-P. Laffitte and C. Denoël, *Trésors carolingiens: livres manuscrits de Charlemagne à Charles le Chauve* (Paris: Bibliothèque nationale de France, 2007), no. 8. Facsimile: F. Crivello, E. König, C. Denoël and P. Orth (eds), *Das Godescalc-Evangelistar* (Simbach/Inn: Faksimile Verlag, 2018).

44 Épernay, Bibliothèque municipale, MS 1: W. Koehler and F. Mütherich, *Die Schule von Reims*, Die karolingischen Miniaturen 6, 1 (Berlin: Deutscher Verlag für Kunstwissenschaft, 1994), pp. 73–84; K. van der Horst and W. Wüstefeld, *The Utrecht Psalter in Medieval Art* (Utrecht: HES, 1996), no. 6; Laffitte and Denoël, *Trésors carolingiens*, no. 41; Nees, *Frankish Manuscripts*, no. 56. The interlace in the lower terminals of the 'N' of 'Initium' is poorly balanced (fol. 61r: Koehler and Mütherich, *Die Schule von Reims*, plate 19), as is some of that within the letter-shapes of the 'IN' of 'In principio' (fol. 135r: Laffitte and Denoël, *Trésors carolingiens*, p. 172).

45 Paris, Bibliothèque nationale de France, MS lat. 266: Rand, *Survey*, 1, no. 119, and 2, plate 136; W. Koehler, *Die Schule von Tours*, Die karolingischen Miniaturen I, 3 vols (Berlin: Bruno Cassirer, 1930-33), 1, pp. 260–9; 2, pp. 71–83; and 3, plates 98–105; Laffitte and Denoël, *Trésors carolingiens*, no. 12.

EU authorised representative for GPSR:
Easy Access System Europe, Mustamäe tee 50,
10621 Tallinn, Estonia
gpsr.requests@easproject.com

www.ingramcontent.com/pod-product-compliance
Lightning Source LLC
Chambersburg PA
CBHW081156290426
44108CB00018B/2566